W9-BXB-717

Curtis Moore
and Alan Miller

Japan, Germany, the United States, and the Race for Environmental Technology

With a new Preface

DISCARDED
JENKS LRC
GORDON COLLEGE

Beacon Press
Boston

To my wife, Judith, without whose support I would never have been able to complete this or much else, and to our children, Sarah and Travis. And to former U.S. Senator Robert T. Stafford, a public servant who never forgot the meaning of that term, and his legislative director, the late Vic Maerki, who never let others forget it either.

C.A.M.

TS
171.4
.M66
1994

To Susan and Joanna.

A.S.M.

Beacon Press
25 Beacon Street
Boston, Massachusetts 02108-2892

Beacon Press books
are published under the auspices of
the Unitarian Universalist Association of Congregations.

© 1994 by Curtis Moore and Alan Miller
All rights reserved
Printed in the United States of America

99 98 97 96 95 8 7 6 5 4 3 2

Text design by Diane Levy
Composition by Wilsted & Taylor

Library of Congress Cataloging-in-Publication Data

Moore, Curtis.
Green gold: Japan, Germany, the United States, and the race for environmental technology / Curtis Moore and Alan Miller; with a new preface.
p. cm.
Includes bibliographical references and index.
ISBN 0-8070-8531-6
1. Design, Industrial—Environmental aspects. 2. Industries—Environmental aspects. 3. Green technology. 4. Pollution. I. Miller, Alan S. II. Title.
TS171.4.M66 1995
658.4'08—dc20 95-21420

JENKS L.R.C.
GORDON COLLEGE
255 GRAPEVINE RD.
WENHAM, MA 01984-1895

Contents

Preface to the Paperback Edition

Green Gold is about change, a word heard with increasing frequency in the United States. Yet what is today billed as "change" isn't. It is instead the century-old status quo. Newly elected officials in Washington—especially Republicans, but also Democrats—are acting to prevent genuine change and assure that the United States ends the twentieth century as it began it, using the same fuels to power the same engines to perform the same tasks in essentially the same way as in the days of Thomas Edison and Henry Ford. The rest of the world is changing, while the United States is standing still—and in the process freezing itself out of the largest, most profitable market in history: namely, the market for "green" technologies that are better, faster, or cheaper because they're cleaner.

Outside the United States, there is little demand for cars that waste four out of every five gallons of gasoline or power plants that dissipate two-thirds of their fuel as useless heat. In other nations, customers are demanding goods that produce more and pollute less. This market is not limited to cars and electricity. Nations from Germany to Thailand are demanding products that range from chlorine-free paper to "smart" energy-saving hotel rooms.

Already many products are intrinsically cleaner and eventually all will be. The nation that produces those cleaner products will be the nation that prospers. New fuels, engines, paints, inks, refrigerants, appliances, and other goods must be developed which, in turn, requires new manufacturing processes—in short, change of the sort that businesses are rarely willing to undertake voluntarily.

Yet in the face of such manifest need for change, support for programs that might yield newer, cheaper, cleaner ways of producing energy is being slashed. Laws that might spur the development of such technologies are faring even worse. The new Republican majority in Congress, accelerating a trend that dates to the early 1980s, has moved to repeal the Clean Air and Water Acts and eviscerate virtually all other environmental statutes enacted in the past quarter century. All these actions have been justified, as they usually are, as necessary to save jobs and produce a prosperous economy.

But in the process these opponents of change have destroyed American jobs rather than preserving them. We believe the jobs lost number, at least, in the tens of thousands and income in the tens of billions. We know, and *Green Gold* demonstrates, that there are technologies now selling throughout the world, generating both jobs and income for Germans, Japanese, Swedes, Canadians, and others, that were invented in America, developed with U.S. taxes, and then sold for pennies on the dollar—if that—because the market for them had ceased to exist here. Many of those products have now traveled full circle and are being sold to Americans.

This foresight on the part of leaders of other nations, and lack of it in the United States, has continued too long to be an aberration. Clearly, these are trends. The United States is moving in one direction while the rest of the industrialized world is moving in precisely the opposite way and the gap is widening. There is no evidence of a change in course. The toll in terms of U.S. jobs and competitiveness is growing, not shrinking.

Green Gold describes these challenges and proposes sensible, workable solutions. The book explains why change—*real* change—is not merely necessary, but inevitable. We hope a realization of this by the readers of this book will be a first step toward reversing course and restoring not just the American environment, but its economy as well.

Acknowledgments

In a book of truly global scope, it is impossible to either adequately or completely acknowledge the assistance of those who have provided help, information, or support. So, we would like first to express our deepest gratitude to those who not only rendered invaluable assistance but were utterly forgotten by the authors at one of the book's most critical stages: namely, the writing of these acknowledgments. Thank you all, first for your help and second for your understanding and forgiveness of our forgetfulness.

The contributions of many others are, however, bold in our memory. In California: the staff and board members of the South Coast Air Quality Management District, especially Hank Wedaa, Larry Berg, Norton Younglove, Henry Morgan, Alan Lloyd, and Jon Leonard; the cheerful folks at "The Gas Company," Ann Smith, Pete Jonkers, and Laurie Kasper Gwynn; in Sacramento, Jan Sharpless, Chuck Imbrecht, Jim Boyd, Bill Lockett, Tom Cackette, Peter Asmus, V. John White, Ken Smith, and Paul Fadelli; in San Francisco, Gigi Coe and Hap Boyd, as well as Hap's frequent sidekick, Jim Caldwell.

In Germany: Axel Friedrich and the many dedicated public servants at the Unweltbundesamt in Berlin, as well as Lutz Wicke and Helmut Weidner.

In Japan: the many workers and managers at the Tokyo Electric

Power Company, especially the redoubtable Ken-Ichi Karasawa and his colleagues Yoshihiro Kageyama and Akira Tanabe; Rioji Anahara, formerly of Fuji Electric; Yoichi Maeda of the Mitsubishi Corporation; Toshiaki Okubo of the U.S. Embassy; and, even though they either are or were in the United States rather than Japan, Takefumi Fukumizu and Gordon Epstein. Many officials at the Environment Agency of Japan provided significant assistance, but we note especially the help of Koya Ishino, Yasuhiro Shimizu, and Hirohiko Nishikubo. Miwako Kurosaka and Haruki Tsuchiya also contributed greatly to our understanding of Japanese energy and environmental policy. We count Tetsuo Nishide of MITI and the Japan External Trade Organization and Koichiro Fujikura of the University of Tokyo as friends as well as advisors.

In Washington (and, for those more fortunate, outside it): Peter Blair, Rosina Bierbaum, Bob Friedman, and Wendell Fletcher, all of the Office of Technology Assessment; Bob Rose, Joe Povey, Eugene Zeltman, James MacKenzie, Jon Fisher, Curt Suplee, Phil Shabecoff, John Hoffman, Douglas Cogan, Bud Ward, Henry Kelly, and David Goldstein.

Special thanks are due our colleagues at the Center for Global Change, Dilip Ahuja, Susan Conbere, Chris Fox, Andrew Hoerner, Irving Mintzer, Frank Muller, and Harvey Sachs, whose insights, criticisms, and emotional support were invaluable. Another colleague, Marybeth Shea, merits special mention for editorial assistance and for the preparation of several summaries and monographs that evolved in the course of our writing this book. Special thanks as well to Lorraine Frankfurt, who typed not only this entire manuscript—many times over—but much else besides. Also, thanks especially to our editors, Deanne Urmy and Andy Hrycyna. They made this book possible, first by believing in the project, then by saving us from our own hastily or poorly written words (though probably not often enough).

Finally, there are those who contributed financially to this project and without whose help it would have been difficult, perhaps impossible, to see it through. They include the Center for Global Partnership, the Joyce Foundation, the W. Alton Jones Foundation, and the Energy Foundation.

Introduction

Propelled by a series of interlocking environmental imperatives, the world is rapidly moving into a new technological era, a second Industrial Revolution every bit as important and long-lasting as that of two centuries ago. Call these imperatives environmental protection, energy conservation, consumer demand, or something else entirely, they are very real—and, more importantly, they are changing the way business is done throughout the world, yielding a profusion of new products that are better at what they do precisely because they are, for lack of a better term, cleaner.

New light bulbs, for example, brighten rooms with 80 percent less energy and pollution than old ones, while saving money. Cars travel two or three times as far on a gallon of gasoline as did cars in past decades, slashing the automobile's contribution to air pollution by up to 90 percent while protecting their occupants better. Electricity can be generated with no pollution whatsoever, and at the same or less cost as in the past.

Had this revolution occurred a dozen years ago, the United States would have reaped the benefits. Largely because of massive spending in the defense and space areas, the United States was the first to develop many of these space-age technologies, but through its lax environmental and industrial policies of the 1980s allowed them to fall into the hands of Japan, Germany, and other competing nations. Now, as nations and consumers of the world demand goods that are cleaner, as well as better, faster, and cheaper, the United States is falling behind.

In Germany, Japan, and many other industrial nations, environmental protection has evolved into a strategy for enhancing economic competitiveness. The Japanese, in search of what one official terms "limitless profits," see a golden future for the nation that captures the market in these new, "green" technologies. In Germany, government and business alike are driven by a public that is relentlessly protective of the environment, but there too, officials say, "What we're doing here is economic policy, not environmental policy."

This vision of the future is shared by nations as different as Singapore and Sweden. But in the United States, decisions are too often driven by the outmoded and false view that the environment can be protected only at the cost of the economy, when the truth is precisely the opposite. If that trade-off ever in fact existed, it was in the days when industries responded to environmental regulations by simply buying a device that screwed onto the end of a smokestack or sewer pipe. Today, when air and water pollution must be slashed at the same time that trash is recycled and products are repackaged, a single end-of-the-pipe device is no longer an option.

Consider, for example, the remarkable story of El Pollo Loco, which traces its roots to a small roadside stand opened in 1975 by Jose and Pancho Ochoa on the west coast of Mexico. There they char-broiled chickens over an open flame, producing clouds of billowing, fragrant smoke—and chicken that sold like salsa and margaritas. The Ochoas' single stand soon became several. Acquired by a U.S. fast-food company and renamed El Pollo Loco, they grew to a chain of 190 restaurants, almost all in Los Angeles, still selling the Ochoas' moist, sweet—and smoky—poultry. It was the smoke that attracted local air pollution regulators, who ordered the chain to cut its air pollution, first by 60 percent, then to a limit of only one pound of smoke a day for each outlet.

Like most companies initially confronted with such stringent requirements, the officers of El Pollo Loco sought an add-on pollution control system good enough to comply and cheap enough to be economically viable. They couldn't find one. So El Pollo Loco, in the words of its chief operating officer, James C. Verney, "got off the roof and into the kitchen," prowling the smoky grills, poking chicken, and trying to find a solution. "We knew we had to eliminate the smoke and grease emissions and at the same time preserve the unique taste of the chicken," recalled Verney in 1992. The answer that El Pollo Loco eventually hit on was simple: the chicken was hung vertically on a chain-driven conveyor, not unlike sides of beef in a warehouse. Passed between red-hot grids, the chicken cooked in half the time, with all the grease falling into catch basins below. The taste was not merely the same but even better, because the chicken was moister and more uniformly cooked. Labor, energy, and cooking costs were all reduced, smoke and grease fires were eliminated, and fewer chickens were wasted.

Looking back on it, Verney said that the pollution control requirement was really "an economic opportunity for us, because it compelled us to be more creative and innovative than we would have been without the clean-air challenge." El Pollo Loco's solution not only reduced pollution, but lowered labor costs, cut energy consumption, lessened cooking time, and increased product consistency and quality control. Staying "on the roof" would have solved only one of those problems.

The El Pollo Loco case is emblematic of opportunities that extend to all parts of the economy, opportunities to become more competitive by becoming more environmentally sound. On its surface, this outcome seems counterintuitive and surprising, because conventional wisdom holds that there is invariably tension between environmental protection and economic efficiency. The proposition that cleaner air and water mean higher costs and fewer jobs is accepted almost universally by economists, businesses, and government officials in the United States. There is reason for this. When crude environmental controls—the kinds that industries are quick to buy as add-on control devices in order to avoid the nuisance (and risk) of examining their products and production lines—are employed, lower productivity may indeed be the result. But when cleaner air and water are achieved by developing a superior product or production line, the results can be—*will be*—lower costs and more jobs.

Still, companies understandably resist what El Pollo Loco's Verney describes as "getting into the kitchen," fearing that the results of a wrong choice can be catastrophic. Few companies enjoying reasonable profits are willing to expose themselves voluntarily to the risks entailed by tinkering with products and production lines, because the risks of failure usually outweigh the rewards of success in the minds of executives with their eyes on the next quarter's profits. Indeed, in many respects environmental cleanliness is to the 1990s what quality control was to the 1960s; the new trend is only beginning to emerge and in some places is yet to be recognized. Increasingly, however, most of America's industrial competitors (and a few of its own companies) are pursuing environmental protection, as they did quality control, because products are better, production costs are lower, and customers demand it.

In fact, of the world's industrialized nations only the United States has yet to fully appreciate the lasting significance of the change being wrought by burgeoning environmental concerns. Germany and Japan, as well as virtually all the rest of America's primary industrial competitors, have adopted a wide range of policies designed to coax or compel the development and commercialization of technologies, practices, and industries that do their jobs as well or better than in the past while producing less pollution. These pressures not only upgrade the efficiency of domestic industries, but also develop goods and services that can be marketed elsewhere; they have produced automobile, steel, electric utility, home appliance, and a wide range of other industries that are either more efficient than their American counterparts or produce products that are. The Japanese and Germans—as well as the citizens of most other industrialized nations—not only produce more steel with less energy and pollution than Americans do, but wash clothes, heat homes, and shower their bodies with less waste and pollution as well.

In Germany, the residues of air pollution control systems are used in the construction of homes and factories. In Japan, energy is wrung out of every pound of coal or gallon of oil—and pollution reduced accordingly—by successions of conservation campaigns at factories. Belgium is building massive wind farms, deploying zero-polluting and cheaper-than-coal wind turbines to generate electricity, while Sweden has developed the world's most energy-efficient buildings. All of these goods and services can be sold elsewhere. Belgium's wind turbines, for example, compete in the United States head to head with the best this country has to offer. Germany has exported its pollution-into-homes process to England, while Japan schools foreign engineers in the use of its products so they'll go home and order them.

The demand for such goods and services extends to developed and developing nations alike because environmental protection is now an economic necessity, which means that the nation that does it best will prosper the most. This awareness has spurred foreign companies to snap up U.S. firms and patents for environmentally promising technologies. For example, the nation's largest manufacturer of solar photovoltaic cells—devices for making electricity from sunlight—has been purchased by the German conglomerate Siemens. Similarly, 49.5 percent of one of the most promising new photovoltaic technologies, developed in large part with U.S. government funding, has been bought by the Japanese office-equipment giant Canon.

Companies in Germany and Japan, working with governments which give them incentives through creative industrial policies, are changing the way they think about business and the environment. Important properties are passing out of American hands largely because U.S. businesses and U.S. policymakers still believe that economic progress is made *despite* the costs of environmental protection. There are U.S. businesses that know otherwise. They range from the Global Environment Fund, a $50-million-per-year investment fund, to the $64-billion-per-year telecommunications giant AT & T, which is now selecting its products on the basis of their "environmental soundness" as well as cost, functionality, and quality. Why? Because "[AT & T operating] units must design products with global environmental regulations in mind or find their products barred or rejected by other countries," the company explains.[1]

Still, such companies are the exception, not the rule, because a dozen years of political and business leadership hostile to environmental protection has dissipated the United States' once dominant position in a wide range of technologies that are certain to form the basis of the industrial infrastructure of the twenty-first century. These range from devices for producing electricity from sunlight to advanced jet engines that have the potential to provide nearly zero-polluting energy to utilities and businesses. Many of these technologies were developed in the United States as a result of massive spending on its defense and space programs, but other nations have acquired many of them for pennies on the dollar, or in some cases for nothing at all.

To understand the mechanisms and implications of these losses, it is worth looking at one example in detail. One of the casualties of U.S. indifference during the 1980s was the firm Energy Conversion Devices (ECD) of Troy, Michigan (a suburb of Detroit). ECD, one of the fastest growing of the nation's high-tech companies, is headed by its founder, Stan Ovshinsky.

Now in his seventies, frail but keen-eyed, Ovshinsky is the driving force behind Energy Conversion Devices, a $300-million company that specializes in cutting-edge energy technologies ranging from batteries to solar photovoltaic cells. The holder of over 400 patents in the United States and another 1,100 abroad, ECD has partnered with companies ranging from ARCO to Canon.

Though Ovshinsky's formal education halted after high school, the Akron, Ohio–born inventor has learned from Nobel laureates and industrial tycoons. A quarter of a century ago, operating from a garage on McNichols Street in Detroit, he formed a company to perfect a machine switch using amorphous materials such as glass. After a telephone call to John Bardeen, inventor of the transistor and the first person to receive two Nobel prizes in physics, Ovshinsky became hooked on the seemingly limitless potential of amorphous solids—materials, like glass and rubber, which lack a crystalline structure. By 1992, the garage had grown to a loose-knit sprawl of six buildings in Troy, and the field of amorphous solids, virtually untapped in the 1960s, had yielded fiber-optic cables, Xerox machines, and devices which can turn sunlight into electricity. The last of these is one of Ovshinsky's specialties.

In the fall of 1992, the center of attraction for most visitors to ECD was the world's newest, largest, and most modern assembly line for the production of amorphous silicon photovoltaic (PV) cells. Ovshinsky had pioneered a unique method of manufacturing the cells by, in effect, printing them, using silicon alloys, on stainless steel backing in much the same way that inks are applied to magazine pages or chemical emulsions to photographic film. By 1992, ECD's employees could produce "sun film" by the hundreds of square yards and visions of printing it by the square mile danced in their heads.[2]

The latest effort involved stainless steel sheets rolled like aluminum foil into belts fourteen inches wide and roughly a half-mile long. The machine required was a marvel of American engineering, but what stuck in the craw of some observers was the plant's destination: by year's end, it would be disassembled, crated, and shipped to KVANT, Russia's leading crystalline photovoltaic manufacturing company. While American investors had all but ignored Ovshinsky's technological advances, the Russians—the *anticapitalist* Russians—had not: they had ordered the assembly line from him while the nation was still under Communist rule. Two other, earlier vintage machines were operating in Japan as well. But here in the United States, it was tough sledding.

Like many U.S. companies, ECD remained unsure in the 1990s whether it could survive as an American firm. "You don't know what we went through [in the 1980s] to keep our independence. We withstood it, and it's taken a toll, in a sense, of our personal lives," Ovshinsky said in an interview conducted by the authors with ECD officers in 1992.

The most articulate and outspoken of ECD's officers on this particular day was a physicist. Steven J. Hudgens, ECD vice president, was clearly angry, recalling the company's struggles during the 1980s to find U.S. capital—and its sometimes reluctant involvement with foreign firms. Without a pause, almost without a breath, Hudgens unleashed a bitter criticism of U.S. policies during the 1980s.

> Two things, I think, happened to cause us to be more involved with foreign companies than domestic companies. During that period of the 1980s, foreign companies, and particularly Japanese companies, were particularly aggressive about planning ahead—developing technologies, acquiring licenses to practice technology, forming joint-venture relationships—to develop products that they perceived would be useful throughout the 1990s and into the next century.
>
> At the same time, American companies were cutting back. RCA had an excellent research lab—they are no more. GE's research lab has

> been cut back to nothing—much of the stuff has been sold out. Westinghouse used to be a leader—no more. . . .
>
> Everyone was looking at what the return was going to be this quarter or next quarter. . . . At the same time that was happening, investment capital which would be used to plan and build and put in improvements and in manufacturing and so forth, was expensive because . . . in addition to the competition from this investment in planned capacity and research, there was also competition from the S & L and junk-bond leveraged buyout stuff competing for the same capital. . . . A sort of feeding frenzy went on in the 1980s, in terms of investment, that caused very little money to be put into long-term planning and R & D.
>
> At the same time there was a management philosophy that would not support [long-term R & D] even if the money were available. The Japanese were much more progressive making those long-term investments, much more interested in the technologies that were being developed, and they ended up being, for the most part, our partners and our licensees.
>
> We are a publicly held company. We have got to do something in terms of expanding our business for the sake of our stockholders. We would love to [partner] with American companies, but the opportunities that existed were in Japan. . . . They were there. They had the cash, they had the foresight, they had the plan.

During the 1980s, Ovshinsky had no choice but to sell the fruits of his genius, though he did invent an ingenious method of eating his cake and having it too. Desperate for capital to hire more engineers and build more solar film assembly lines, ECD's officers scoured the United States seeking investors, but in vain. Ovshinsky was left with no choice but to enter into a partnership with the Japanese firm Canon.[3] To retain control, or close to it, Ovshinsky agreed with Canon to form a new company, United Solar Systems Corporation (USSC), with three owners: Canon held 49.5 percent of the stock, ECD held another 49.5 percent, and the remaining 1 percent—and with it control over the corporation's future—was held by Edwin O. Reischauer, President Kennedy's ambassador to Japan. Revered by the Japanese for his role in helping rebuild their nation's shattered economy, Reischauer was also a member of ECD's board of directors and a longtime friend of Ovshinsky's.

"Here was a guy that both entities had the ultimate confidence and respect for," said Hudgens. "Canon is an eight- or ten-billion-dollar company, and ECD is this little entity—how could we have had a fifty-fifty joint venture with a tiger like that? Canon is not a charitable organization. They are aggressive and predatory in the marketplace. They were an excellent partner, but you want to be careful that it doesn't occur to them that they could perhaps eat you along the way."

So Ovshinsky didn't exactly lose his technology. But he lost something important, nevertheless. "What we lost was independence," said Hudgens. "That is not ECD anymore—that's USSC. That pays attention to the interests of our Japanese partners as well as our own. It's no longer an ECD enterprise. Although it's an American company, it's an American company with interests that are not exclusively American, because 49.99 percent is Japanese."

Yet many of ECD's U.S.–based competitors failed to survive at all, much less retain nearly half of their property. They either went bankrupt completely or, faced with neither a U.S. market nor U.S. buyers, simply sold their technologies to foreigners.

Space-Age Environmental Technologies

The origin of many of the technologies most closely linked with environmental issues can be traced to October 4, 1957, the day the Soviet Union vaulted past the United States to gain in a lead in the "race to conquer space" by launching *Sputnik*. American embarrassment deteriorated into humiliation during the months that followed, as one U.S. rocket after another exploded on or just over its launching pad. The space race was placed on a wartime footing, and vast sums were spent to unleash a torrent of innovation.

Technologies later called environmental (directly or indirectly) included:

- solar photovoltaic cells, developed to make electricity from sunlight in order to power the high-flying satellites as they traveled through space;
- fuel cells to convert hydrogen and oxygen—carried on board the space rockets and shuttles for other purposes—

into pure water and electricity for the Apollo and Gemini manned space programs;

- aircraft-derivative turbines, known simply as jet engines to most of us, to hurl war machines through the air at two, three, or even four times the speed of sound.

By the mid-1970s, Americans had walked on the moon, circled the Earth, and made it possible for voices and images to flash at the speed of light from any given point on the globe to another. But technological momentum had slowed, and many inventions were left to molder on the shelves of American industry.

Then came another shock, triggered in the Persian Gulf. In October 1973 the Arab oil-producing nations choked off the flow of petroleum to the world, demonstrating that there were other ways of measuring international power. Military might and domination of space was one—oil was another.

The industrialized nations responded in much the same way that the United States had reacted to the steady beep of *Sputnik*. America, Japan, Germany, France, Sweden, and other users of oil mounted massive programs to achieve "energy independence," as it was then called by President Richard M. Nixon. Use of coal boomed. Corporate wildcatters, tantalized by the fortunes to be made in exploiting oil-saturated rocks and sands, created million-dollar boomtowns in the arid American West, overwhelming tiny communities with picturesque names like Parachute and Gillette.

The natural gas and electric utility industries started pumping money into fuel cells—devices that chemically convert fuel into electricity—in attempts to double or even triple the amount of electricity generated from a given amount of fuel. Oil companies—Exxon, Amoco, Standard Oil, ARCO, and Shell, to name a few—created new divisions devoted to developing ground-based systems for generating electricity from sunlight. Jet engine manufacturers, principally General Electric, launched programs to use their jet engines at pulp and paper mills, chemical plants, and other facilities to generate both electricity for sale to local utilities and the steam used for the manufacturing operations themselves.

There was pressure from the coal, oil, auto, steel, utility, and other industries to loosen environmental controls that had been adopted in the early 1970s, but these measures remained largely intact and stringent. As a result, carmakers, power engineering firms, and others began developing add-on methods of controlling air pollution. Catalytic converters—devices that when clamped onto tailpipes can eliminate 90 percent of car pollution—became standard equipment in the United States. Clean air mandates combined with new federal safety and fuel-efficiency requirements to spawn a new fleet of cars. Within fifteen years, America's cars could travel two or three times further per gallon of gasoline while reducing air pollution and saving lives through improved safety engineering.

These forces combined to produce an unprecedented level of U.S. dominance in virtually every environmental technology. By 1988, when the world suddenly awakened to the threat of global warming and an environmental consciousness gripped consumers and governments alike, the United States should have been in a position to establish itself as the unrivaled industrial power of the twenty-first century, for it had developed and built the world's best gas turbines, wind turbines, solar cells, fuel cells, and antipollution scrubbers—everything from on-board car computers to advanced light bulbs was housed in America's technological arsenal.

Yet when the time arrived to beat those technological swords into plowshares, most were missing. The Japanese had been allowed to become the world's largest manufacturer of solar cells. The United States still occupied second place, but America's largest solar cell factory was owned by a German firm. The patents on one U.S. fuel cell technology were sold to a Japanese company, Fuji Electric, with the approval of the U.S. government; those on another were allowed to expire, falling into the hands of a Canadian company, Ballard Power Systems. One major fuel cell manufacturer had started an assembly line, but it was owned in significant part by a Japanese firm, Toshiba.

The world's largest manufacturer of catalytic converters—pollution control devices first mandated by the United States—was a British firm, Johnson-Matthey. The largest makers of high-efficiency lighting systems, developed in large part with funding from the U.S. Department of Energy, were Japanese, Dutch, and

German. Hundreds of selective catalytic reduction systems, which eliminate pollutants that cause smog and acid rain, were installed on power plants and factories throughout the world. The catalyst at the heart of these systems had been developed by Corning Incorporated in New York, but sold when a market for it couldn't be found in the United States.

In the 1970s, the United States dominated the market in scrubbers, add-on pollution control systems for removing sulfur dioxide from power plant smokestacks. Today, Japan and Germany are the world's technological leaders in the field. It was the United States that provided the incentives to build and operate the world's largest concentration of devices for making electricity by using the sun's heat, generating enough electricity for a city the size of San Francisco. Yet today the company that built those power stations is bankrupt, and it is Germans, not Americans, who talk of exploiting the technology throughout the world's sunbelt. The United States developed and demonstrated the "Cool Water" technology to burn coal more cleanly, yet today it is being improved by the Japanese and the Dutch. America developed nuclear power—first for bombs, then to power submarines, then to make electricity commercially—but today it is the French whose concentration of reactors is the world's largest, the Japanese whose safety record is best, and the Swedes, Japanese, and Germans whose new designs are the most advanced.

The loss of U.S. dominance in new technologies is graphically illustrated by a power plant being built in the rolling countryside near Doswell, Virginia. There, only a two-hour drive from the Capitol, a new gas-fired power plant is being constructed to provide enough electricity for a city about the size of Richmond, Virginia. The plant's owner is Mitsubishi. The manufacturer of its steam turbines is Asea Brown Boveri, a Swedish-German-Swiss conglomerate. The manufacturer of the gas turbines? Siemens, a German conglomerate. The manufacturer of the selective catalytic reduction systems for controlling air pollution? Mitsubishi.

There are some bright spots in the U.S. position. General Electric, for example, remains the world's leading manufacturer of combined-cycle turbines. GE and Pratt & Whitney retain their positions as two of the world's three makers of compact, lightweight, and low-polluting aircraft-derivative turbines. International Fuel Cell Corporation—parent company of one of the world's two assembly-line manufacturers of fuel cells—may be owned in part by Toshiba, but the controlling interest is still in American hands. These examples, however, are increasingly the exception rather than the rule.

Foreign manufacturers have continued to become stronger in the United States. They look to the American market for additional growth opportunities. Most foreign companies started doing business in the United States as small enterprises importing a few pieces of equipment here and there. As they became better acquainted with this market and their competition, they saw opportunities to buy small firms or to form joint ventures. For joint ventures, weaker companies that needed help were usually chosen; in many cases, it was not long before the foreign partner took over the U.S. organization and created its own operating entity in the United States.[4]

An analysis of the electrical equipment and power systems industry clearly shows that U.S. electrical equipment manufacturing is in decline; this is also true of the mechanical and nuclear components sectors that make up the power system. Electronic controls and monitoring, together with a limited amount of so-called clean coal technology, are a few of the areas in which U.S. companies are holding their own.[5]

An assessment published by the U.S. National Academy of Engineering is bleak, finding that U.S. leadership has declined or probably will be declining in the production of steam turbines, steam generators, nuclear reactors (especially the so-called inherently safe versions, which are designed to be virtually failure-proof), and pressurized fluidized-bed combustion systems for burning coal more cleanly.[6]

Foreign companies already claim about 25 percent of the U.S. electrical equipment market. America's political leaders are counting on the ability of U.S. manufacturers to use the remaining 75-percent share as a springboard into the global market. But if the construction of power plants by foreign companies—Mitsubishi's Doswell facility, for example—becomes a trend, U.S. manufacturers will forfeit larger and larger shares of the domestic market to foreign firms and as a result lessen their ability to compete globally. The simple truth is that the United States cannot sell what it does not produce.

There are those who remain untroubled by the massive drain of environmental and related technologies. Their view is that if products, whether they're plastic whistles or fuels cells, are made in the United States, then jobs are provided for Americans, so the ownership of the company is irrelevant.

Such a view ignores the importance of the ownership of the technology itself to the continued viability of America as an industrial power and hence to our prosperity. Jobs in foreign-owned companies in the United States may for the moment be American, but the profits are not and never will be. Instead of profits flowing to investors in, say, St. Louis, they—and the future prosperity they assure—run to investors in Yokohama or Munich. It is invention that creates true wealth, and true wealth that creates lasting jobs. A United States that loses control over the true source of jobs—the wealth created by ingenuity and innovation—will be a nation that has lost control over its own destiny and independence.

This book examines these technologies, the policies of other nations, the environmental imperatives, and the possible future for this country should our leaders grasp the opportunity. We will first discuss the new realities in detail, paying special attention to the policies of Germany and Japan because these have been the first nations to fully appreciate the competitive situation. Next, we will examine why the United States has fallen so far so fast, dissipating a once commanding position of technological and environmental leadership. Finally, we will explore the potential for an American resurgence exploiting the massive U.S. reserves of energy and ingenuity.

What is at stake is not a splinter market, but virtually all goods manufactured for almost all purposes. Producers of goods ranging from yogurt cartons to cars must increasingly respond to the new environmental imperatives. In a global economy, no nation can successfully isolate itself from the policies and demands of others. A nation that attempts to do so will soon find itself with a dwindling share of the international trade in manufactured goods and, as a consequence, with a shrinking standard of living.

American leaders must awaken to the competitive threat—and challenge—posed by environmental imperatives, for seldom has the nation been presented with a better opportunity to reestablish its industrial and commercial leadership. The United States remains not only a powerful technological force, but a fabulously wealthy nation in terms of its reserves of solar, wind, and other nonpolluting energies. There is no good reason that limitless profits should flow to Japanese and German investors for technologies that were developed with American sacrifice and dollars.

PART I

NEW REALITIES— AND REALISTS

Chapter One

Germany's Miracle

What we are doing here is economic policy, not environmental policy.
—EDDA MÜLLER, German Environment Ministry

Heaped near the railroad tracks at Knauf Gypsum, Germany's largest construction materials company, lies what would once have been a good share of the nation's air pollution. Soon it will be made into homes.[1]

The gypsum briquettes, hard as rocks and roughly the size and color of eggs, lie there, heaped, through the winter's bitter cold and snow. When spring arrives, the piles begin shrinking as the briquettes are pulverized, then mixed with water to make a thin paste that's spread between sheets of heavy kraft paper and dried to form a rock sandwich called wallboard, sheetrock, or gypsum board. As the construction season begins with the fading of winter, the board is shipped to building sites across the nation, where it is nailed to studs to form the walls and ceilings of offices and bedrooms, closets and boardrooms.

Elsewhere in the world, when coal is burned to generate electricity, prodigious amounts of pollutants pour into the air, including sulfur dioxide, which causes acid rain. Some nations, but not many, require modest controls over this pollution. If the controls are stringent, compliance usually takes the form of scrubbers, devices that remove the sulfur dioxide by spraying the exhaust with a watery mist containing limestone. The pollution-limestone reaction produces a watery sludge that is, almost everywhere but in Germany, simply dumped on the ground or into pits. But in Germany, all power plants are equipped with pollution controls and the sludge can't be dumped because the law prohibits it. Such waste must be put to some use, leaving German power plants with two options: develop some means other than scrubbers to eliminate the air pollution, or find a way to use the scrubber sludge—like using it to make gypsum board.

German industry has done both, yielding two simultaneous streams of innovation, one aimed at developing pollution control systems better than scrubbers, the other at perfecting better ways to use scrubber waste. Both streams not only help make the German economy itself more efficient, but also encourage the development of products that can be sold in the world market, boosting employment and income at home.

The Knauf Gypsum program is so successful that the company opened a new plant in 1990 at Sittingborne-on-Thames. From there, sheetrock is shipped throughout the United Kingdom so English homes and offices can also be built from German air pollution. The process of turning pollutants into building materials for homes and offices is one on which Knauf makes a substantial profit—almost certainly in the tens of millions of dollars per year, though officials decline to reveal specifics.

This process of making homes from pollution is emblematic of the innovations that have sprung up in Germany as the nation has leapt into the forefront of the global environmental movement. As pollution in all its forms is eliminated, Germany is adding more muscle to an already lean, efficient economy. Its industries and products are becoming more competitive, and its advances are selling throughout the world. The pollution-to-homes process illustrates how environmental concerns have stimulated innovation.

That stringent environmental safeguards enhance competitiveness is increasingly taken for granted in Germany, for from Bonn to Berlin, Germany's citizens, businesses, and government have concluded that the relationship between a robust economy and a safe environment is mutually reinforcing. Spurred first by concern over its environment, then by the prospect of stealing a competitive edge over its industrial rivals for global markets, Germany has vaulted into the vanguard of global environmental leadership in the space of ten years.

Together with Japan, Germany has become a preeminent leader in a wide range of environmental goods and services, but the two nations have decidedly different motivations. Japan's have often been strictly commercial. Many of its advances were developed primarily to serve environmentally sensitive export markets, especially those of the United States in the 1970s and Europe in the 1980s and 1990s; others resulted from efforts to curb Japan's reliance on imported fuels while reducing air pollution. With the rapid rise in concern over global warming, Japanese industrial and government officials see the opportunity to serve not just segments but the entire world economy. Whether global warming and the many other burgeoning environmental concerns are realistic or not, they create a market demand which the Japanese are intent on meeting.

Germany, by contrast, was and very much still is motivated more directly by an intent to protect the environment. Officials and the public there have concluded, for example, that global warming is a genuine threat to human survival and believe that sooner or later other nations will reach the same conclusion. When that happens, the demand for goods and services capable of meeting the threat will intensify and if Germany has developed them it will prosper from the resulting sales.

These differences in emphasis between Germany and Japan are linked to different approaches to policy and industrial development. Japan continues to rely on close collaboration between government and industry and on collective effort. Germany tends to announce environmental requirements, then allow industries to make their own choices about the best means of complying. In a remarkably large number of cases, the two approaches yield similar results. Consider, for example, the following:

Solar photovoltaics. Japan is now the world's leader in production of solar photovoltaics, which are devices for producing electricity from sunlight. The United States has fallen to second position, neck-and-neck with Germany, which is rising fast—so fast, in fact, that in the words of the former head of the U.S. program, "there's no doubt" that Germany will soon surpass the United States.

High-efficiency turbines. Although General Electric remains the world's leader in the manufacture of these machines for producing electricity, the German giant Siemens has produced what are (at least for the moment) the world's least-polluting and most highly efficient turbines. Two such turbines have already been installed in Delaware at a Delmarva Power and Light facility, while others were purchased by Mitsubishi for its Doswell, Virginia, plant. Siemens is pushing hard to close sales in California as well. Japan, which long relied on U.S. manufacturers, now boasts its own companies and has launched efforts to develop a super-efficient "ultra-supercritical" turbine.

Add-on pollution controls. When Germany embarked on its crash program to slash the power plant pollutants that cause acid rain by 90 percent within a six-year period, it was forced to buy technology from the world's leading producer, Japan. German firms have improved that technology and are now aggressively marketing their own versions, challenging Japanese supremacy. The United States, once the world's leader, is a poor third.

Light bulbs. One of the world's leading manufacturers of compact, high-efficiency fluorescent light bulbs and ballasts (which stabilize electric current and energize the light-producing gases with which the bulbs are filled) is Siemens, the German conglomerate. Another is Panasonic of Japan, though these devices were in large measure initially developed with funding from the U.S. Department of Energy.

Motor vehicles. Both Germany and Japan are on the cutting edge in developing zero-polluting electric and hydrogen vehicles, especially cars. BMW and Mercedes have both fielded hydrogen-fueled test vehicles, as has Mazda. Similarly, carmakers in Germany have developed battery-powered cars, as have their Japanese counterparts. In Japan, even the electric utilities see a potential market: Tokyo Electric Power Company—the world's largest private utility, with about 26 million customers—has built an advanced, sleek, and lightweight battery-powered prototype called the IZA. (U.S. manufacturers have also developed electric vehicles, but GM's version, the Impact, has been shelved, while Chrysler's TEVan has been disappointing. Although American carmakers are collaborating to develop batteries, they typically use the relatively limited range and long recharge times of today's versions as arguments against electric vehicles and in favor of conventional gasoline versions.)

For short- and medium-term increases in fuel economy, German manufacturers are focusing their efforts on squeezing maximum mileage and minimum pollution from the diesel engine. International consultant Michael Walsh believes that by the century's end Germany could have a fleet of clean, diesel-powered vehicles averaging 40 to 50 miles per gallon, compared to the 25 to 30 mpg of North American cars, if current trends continue. Japanese carmakers have focused on accelerating the development of high-efficiency "lean-burn" engines. U.S. companies have done neither.

Household appliances. Although German appliances have historically been smaller and more energy-efficient than their North American counterparts, they're becoming even more so. New German washing machines, for example, have computer microchips that sense the weight of a load, metering soap and water accordingly—something that is "still a distant vision" in the United States, according to one expert. Japan has long specialized in compact, highly efficient appliances. U.S. companies may catch up, however, due to minimum efficiency standards mandated by Congress in 1987.

More than any other nation, Germany is betting that its environmental regulations (far and away the world's most stringent) will foster the development of technologies that will form the foundation of the world's industrial infrastructure in the next century. Germany has shown how environmental protection through regulation enhances product development, how technological innovation produces domestic jobs and international competitiveness, and how competitiveness in turn enhances the domestic economy and creates jobs.

This focus is consistent with Germany's longstanding and remarkably successful policy of building competitiveness through technological innovation. German companies are world-class producers of high-quality goods ranging from cars to coffeepots—and the trade figures show it. In 1990, for example, Japan exported $282 billion in manufactured goods, while the U.S. figure was slightly higher at $287 billion, but far and away the preeminent exporter was Germany, with $386 billion—28 percent more than the United States.[2] Much of that trade is at least partly related to the environment, although it's difficult to ascertain exactly how much.

German pollution control technology is now being constructed and installed at power plants in Denmark, the Netherlands, Turkey, and Canada, to name but a few. Officials expect that the competitive advantage will expand as equipment now being developed to comply with more recent regulations is brought to market. This will include heat exchangers, gas turbines, steam motors, and CFC-free absorption air conditioners. Since 1980, the number of environmental companies has quadrupled, rising from about 1,000 to roughly 4,000.[3] Aggregate statistics are almost impossible to obtain, but a wealth of anecdotal data indicates that Germany's share of the world export market is expanding due to the superior environmental performance of its goods.

Consider, for example, the pollution-from-homes process—which, like so many others, was originally developed in the United States. The first commercial plant using this system was installed at the Cholla I station in Arizona, which began operating in 1973. In 1980, the system was exported to Germany, where it was initially marketed by Knauf-Research Cotrell (KRC), a subsidiary of its U.S. developer and Knauf. Rapidly improved in response to the German air pollution and waste requirements, the technology was acquired in October 1986 by the Salzgitter Group and is now marketed throughout the world by one or more of its subsidiary companies. One of the places where the system has been sold is New Brunswick, Canada, where the 450-megawatt coal-fired Belledune power station was put into operation in 1993. The production of market-grade gypsum "was a fundamental requirement" contained in the specifications for the Belledune plant, in the words of an executive of New Brunswick Power, because it not only solved waste disposal problems but was less expensive than competing systems.[4] Thus a North American innovation traveled to Europe and back again in the space of twenty years—although the profits are being made by the Germans—and it is selling globally because it is, quite simply, better.

Although surprisingly few business leaders and economists in the United States are aware of the details of Germany's chicken-and-egg environmental/economic policies, many agree with their fundamental premise. Roger Gale is a former senior official at both the U.S. Department of Energy and the U.S. Environmental Protection Agency who now advises foreign and domestic utilities on the competitive implications of environmental regulations. He says investments in new technologies of the sort being deployed in Germany "accomplish the twin goals of improving environmental quality while improving competitiveness."[5]

"Gains in efficiency from investment in new technologies and services will provide a huge long-term competitive advantage," Gale continues. Economic efficiencies aimed at cleaning up the environment are expected to become the basis of German economic competitiveness over the next century. "Any country that does not emulate Germany's strategy will be at a competitive disadvantage in ten or twenty years," echoes Konrad von Moltke, a senior consultant to the World Wildlife Fund who has written extensively regarding Germany's new policies and is one of only a handful of Americans familiar with the full range of Germany's ambitious programs. "Some nations—Germany, for example—understand this," he added, "while others don't."[6]

In 1980, the positions of the United States and Germany were almost exactly the opposite. The United States had been the first nation to adopt stringent air and water pollution control laws, while Germany resisted them. When the United States banned most spray-can uses of the ozone-destroying CFCs, Germany led European resistance to an extension of the prohibition.

Ironically, this change in German policy occurred under the leadership of a politician widely regarded as among the world's most conservative, Chancellor Helmut Kohl. Although he was often compared to Ronald Reagan during the early 1980s, Kohl took a decidedly different attitude toward protecting the environment than did the U.S. president. Reagan, not entirely in jest, blamed "killer trees" for air pollution, and systematically set about relaxing controls on U.S. industries throughout his eight-year presidency. Kohl, in contrast, pushed environmental regulation for eight years, then, in March 1989 (after Reagan had retired to California), Kohl told a gathering at The Hague that the world was "racing against time." Efforts to protect the environment, he said, had to be "accelerated [and] given new impulses."

Some observers do question whether Germany can achieve its ambitious environmental goals while maintaining a standard of living and a social welfare system that are among the world's most generous—all while undertaking the mammoth costs of reunifying the former East and West Germanies. Gasoline prices have already been hiked by roughly 45¢ a gallon to pay help pay for the estimated $128-billion cost of bringing the former East Germany up to the West's stringent environmental standards.[7] New taxes on toxic wastes, carbon dioxide, and other pollutants will help raise money for these and other reunification efforts while fostering environmental protection. Some industries have begun to resist. For example, when the government proposed yet another turn of the environmental screw in late 1991, the chairman of Hoechst, Germany's largest chemical company, complained bitterly, saying the government "had lost all sense of proportion." Wolfgang Hilger said stringent requirements had already forced Hoechst to halt production of some dyes and chemicals at a cost of $60 million and 90 jobs, and that another 160 jobs and $48 million were threatened. Chemical firms, he complained, "now have to study each new legislative proposal to see whether we can still afford to invest in Germany."[8]

Notwithstanding such objections and despite a bad economy worsened by the formidable burdens imposed by reunification, Germany's government and public, as well as most of its industrial leaders, remain committed to the propositions that environmental protection is essential and that the technological innovation stimulated by the world's most stringent environmental requirements will, over the long term, strengthen their national productivity and competitiveness.[9]

"What we are doing here is economic policy, not environmental policy," as Edda Müller, chief aide to Germany's minister for the environment, says. Still, Germany's initial enthusiasm for environmental protection had little to do with stimulating the economy, for before the early 1980s Germany's environmental protection program could fairly be described as relaxed at best. Prior to 1983, for example, the acceptable level for atmospheric concentrations of sulfur dioxide, the primary cause of acid rain, was over twice the World Health Organization's recommended level.[10] Tall smokestacks—later to come into widespread use in the United States—were used to dispense pollutants over wide areas as the main mechanism of reducing local air pollution.[11] An automobile pollution control program was virtually nonexistent.

This all changed abruptly in the early 1980s. The change can be explained in a single word—*Waldsterben*. Translated literally, it means "forest death," a collapse of entire forest ecosystems caused by air pollution. Needles and leaves yellow, trees sicken imperceptibly at first, then suddenly, in one or two seasons, they're dead. Initial reports about damage to Germany's forests were continually revised upwards during the early 1980s, with researchers concluding in 1985 that 50 percent of the nation's trees were injured.[12]

In Germany, a nation with an almost mythic connection to its forests—from the fabled Black Forest of Hansel and Gretel to the graceful lindens of Berlin—*Waldsterben* had a galvanizing effect. The German people pushed their politicians toward tighter and tighter controls. Then, as momentum began to subside, came the explosion and near-meltdown in 1986 of the Soviet Union's Chernobyl nuclear power plant. As rumors of two-headed calves and human birth defects swept the nation, they fueled a resurgence of environmentalism. Hard on Chernobyl's heels came news of the Antarctic ozone "hole," caused by chlorofluorocarbons, and of rising concerns about global warming. With each successive shock German resolve stiffened. Today there is virtually no field of environmental protection in which Germany does not stand out.

The first of the government's responses to voter alarm was a twofold program aimed at eliminating the cause of *Waldsterben* by slashing pollution from both tailpipes and smokestacks.

First, a stationary-source control program set emission limits so low that they forced adoption by all medium- to large-sized power plants in Germany of state-of-the-art controls—so-called wet scrubbers for SO_2 control and "selective catalytic reduction" (SCR) for the elimination of oxides of nitrogen.[13] Importantly, what the German law required was the achievement of a numerical standard, not the technology itself. Thus polluters who could find a way of meeting the standard that was cheaper, faster, or easier were free to do so—and some did, resulting in a burst of pollution control innovation.[14]

Second, even though the rules of the European Community (EC) barred Germany from unilaterally legislating auto emissions standards, the government adopted tax incentives to encourage the purchase of low-emission vehicles.[15] Among other measures, Germany forgave the national sales, or value-added, tax of 14 percent for cars equipped with catalytic converters.[16] Simultaneously, Germany began pressing for European Community adoption of stringent tailpipe standards. In June 1989, the EC adopted standards for small cars (engines smaller than 1.4 liters) effective beginning in 1992.[17] The EC countries also agreed in principle to enact similarly strict standards for medium- and large-sized cars beginning in 1992.[18]

Germany's new power plant standards were an unusual mix: flexible in some respects, rigid and unyielding in others. They applied to all coal-, oil-, and natural gas–fired boilers larger than 50 megawatts thermal, which is roughly the size required to power three to four thousand U.S. homes. The standards varied somewhat in their stringency, depending upon the plant's size and its fuel, but—unlike U.S. requirements—not upon its age. New and old power plants alike were required to abide by the same requirements, with one exception: if they chose, existing power plants could shut down rather than comply. Those opting for a shutdown were allowed to remain in operation for 30,000 more hours, the rough equivalent of 3.5 years of base-load generation or 10 years of "peaking" (operating a power plant during times of greatest electricity demand, or peaks, such as hot summer afternoons in the United States). In no event, however, could a plant failing to meet the standards operate beyond December 31, 1990—roughly six and a half years after the ordinance's adoption.[19]

The most significant aspect of the new law may not have been its stringent emissions limits, for they have since been surpassed by new technologies. Rather, the revolutionary requirement was the establishment of the principle of *Dynamisierungsklausel*, or dynamic, state-of-the-art regulation.

Under the *Dynamisierungsklausel* principle, regulations for power plants, steel mills, refineries, and other large polluters are subject to constant review to assure that the newest and best pollution control technologies are installed. Although the principle allows for consideration of the costs of control, these are distinctly secondary, with the overriding objective being to promptly achieve maximum possible emissions reductions through the application of state-of-the-art technology.

To determine whether technology is state-of-the-art, regulators look beyond the smokestack to the manufacturing process itself. Thus process changes—even product changes—can be compelled by the principle.

At the Ford auto plant in Cologne, for example, the state-of-the-art principle required the firm to modernize its paint-spray line, cutting pollution by 70 percent. In the process, Ford reduced the cost of painting a car by about sixty dollars—a savings that increases its profits and boosts its competitiveness, if only marginally.

In another application of the principle, at the 4P plastic-film manufacturing and printing plant in Forchheim, where plastic bags for frozen french fries and other foods are printed and stamped by the millions, officials were forced to cut pollution by 70 percent. They installed a recycling system that reclaims up to 90 percent of the plant's solvents. 4P's recycling system recaptures so much spent solvent—previously being emitted as pollution—that it very nearly pays for itself even when chemical prices are low; when solvent prices rise, as they did during the Gulf War, the system begins turning a profit for 4P. At a sister plant, a similar but more efficient system quickly paid for itself, then began saving the company money by reducing solvent purchases.

In recent years, the German government's focus has gradually shifted from requiring add-on pollution controls to developing more efficient manufacturing methods and reducing fuel consumption because, in the words of government expert Axel Friedrich, "the cleanest fuel is the one that's never burned." People and products alike are being systematically shifted to less polluting, more efficient ways of movement.

Inner cities are being systematically closed to auto traffic, while highway, bridge, and other tolls are being raised. The government hopes that bicycle ridership in Germany, already at about one in twenty commuters, will double.[20] Other regulations on the drawing board will reduce pollution by forcing drivers out of gas-guzzling cars and onto public transit. Long-term passes for public transportation, for example, are sold in all of Germany's major cities. A Berlin "green pass" costs about thirty dollars a month, and can be used an unlimited number of times and lent out freely. A comparable pass (sold merely as transportation, not as a "green" pass) in Washington, D.C., costs roughly twice as much.

By systematically encouraging a shift in the transportation of people and goods in this fashion, Germany not only reduces pollution but boosts the overall efficiency of its economy by curbing aggregate transport costs. The less it costs to move people and products, the higher the profits. Moving people on trains, for example, slashes both fuel consumption and air pollution by up to 75 percent compared to using cars, and 90 percent compared to planes.[21] Traffic congestion—and the pollution it generates as cars creep through crowded roadways or idle at stoplights—is also cut because trains occupy only one quarter of the road space of buses and one thirteenth that of cars.[22] Because trains run on electricity generated by Germany's domestic coal, oil imports required to fuel diesel buses or gasoline cars are likewise reduced.

This emphasis on less pollution, high efficiency, and high-speed rail travel has stimulated the development of a new generation of trains in Europe, trains that are now being sold to the United States. Siemens' Inter City Express, or ICE, is competing for the multi-billion-dollar Amtrak contract for high-speed equipment to serve the Boston-Washington rail corridor with the Swedish-Swiss-German conglomerate Asea Brown Boveri. The Spanish firm Talgo is another of the bidders preliminarily approved by Amtrak, together with Fiat Ferroviare of Italy and Bombardier of Canada. The sole U.S. entrant is a Boise, Idaho–based company, Morrison and Knudsen, whose experience in the field is largely limited to the refurbishment of rail cars. Asked to explain the lack of U.S. competitors for the lucrative contract during a period of economic recession, an Amtrak spokesman explained that American investment had historically been directed towards highways and other forms of transportation, largely abandoning rail technology to European and Japanese companies—despite the fact that super-fast, super-smooth magnetic levitation and other technologies had been developed here. "Why would [U.S. companies] bother to build something for which there wasn't a market?" he asked, noting that government tax and public works policies both heavily favored cars and trucks for moving people and goods.[23]

Virtually all European nations are cooperating in this joint effort to shift traffic onto high-speed rail. A network is being planned that will link nearly all of Europe's major cities, ranging from Lisbon and London, in the south and west, to Helsinki and Ankara in the north and east. Trains will travel at nearly 200 miles per hour, eliminating the need for travelers to rely on planes or cars, both of which consume more fuel and generate vastly more air pollution per passenger. The European network will be complemented by new and upgraded national rail systems.[24] After years of stepchild funding, rail systems are receiving massive infusions of money. In Sweden, for example, highway funding is falling while rail spending is rising. All of this will yield much more efficient national systems for transporting both people and goods—and it all followed a commitment by the European nations to cut their carbon dioxide emissions in response to the threat of global warming.

Part of this shift towards less polluting, higher-efficiency ways of doing business has focused on developing sources of energy that produce no pollution whatsoever. The most environmentally elegant of all of Germany's programs may be the aggressive one focused on commercializing what many energy experts consider to be the two perfect fuels, solar energy and hydrogen.

Near the tiny village of Neunberg, about twenty miles from the border with the former Czechoslovakia, is the $38-million Solar Wasserstoff power plant. By utility standards it is minuscule, producing only enough energy to power fifty or sixty U.S. homes by converting solar radiation into electricity and, in turn, producing hydrogen. Still, it generates enough energy to run zero-polluting cars and utterly clean furnaces, and its technology may, perhaps twenty years from now, fuel a future with virtually limitless and absolutely safe energy.

Electricity can be generated from sunlight through the use of photovoltaic panels—devices commonly used on a small scale to power calculators and watches. Solar electricity generating plants are utterly silent and nonpolluting, and once on-line they break down far less often than coal-, oil-, or natural gas–fired power plants.

Still, the price of solar electricity remains a formidable obstacle because it can be up to five times of that made from coal or oil. To reduce costs and gain hands-on experience with engineering details, Germany has launched a number of programs, including the Thousand Roofs program, aimed at installing residential-scale solar-electric panels on roofs throughout the country by providing government purchase subsidies of up to 75 percent. (This wildly successful program, which grew from one thousand roofs to two thousand after expansion to include the former East Germany, was swamped with vastly more requests than it could fill.)

For other uses, hydrogen is an equally perfect fuel: it can be produced by splitting water with electricity, a process called electrolysis. When burned, hydrogen produces only water vapor. While there's little doubt that hydrogen can be used for everything from home furnaces to cars, the infrastructure of pipelines, storage tanks, and the like is lacking, and so is extensive experience in using the highly explosive gas safely.

It was to gain this experience that the governments of Germany and Bavaria[25] teamed with a handful of industrial partners, including BMW and the aerospace giant Messerschmitt-Boelkow-Blohm (MBB), to build Neunberg's Solar Wasserstoff plant. Using its outputs of zero-polluting electricity and hydrogen, German engineers are experimenting with different types of furnaces, cars, storage systems, and other equipment to eliminate the devilish kinks that can spell the difference between success and failure. During off-hours, the excess electricity is sold to the local utility and used to power nearby homes.

In addition, Germany's government is offering funds to Mercedes-Benz and BMW to hasten the development of hydrogen-powered cars and trucks. If the ultimate goal of using solar-derived electricity to decompose water into oxygen and hydrogen is realized, Germany will convert itself to nonpolluting fuels—solar electricity to run homes, shops, and factories, while solar-derived hydrogen fuels the nation's cars, trucks, locomotives, planes, and even ships and submarines.

Peter Hoffmann, editor of the *Hydrogen Newsletter* (published near Washington, D.C.), estimates that Germany is spending about $58 million per year on hydrogen development, compared to roughly $1.4 million in the United States. In Germany, the bulk of these funds is spent on "solid, three-dimensional things, not just generating more paper," he said in an interview. According to Hoffman, "There may be little niches here and there" where other nations are ahead of the Germans in hydrogen technology, "but overall Germany is clearly the leader." Ten years ago, added Hoffmann, the United States was the world's preeminent leader in the development of hydrogen.

As for solar cells, German production is up steadily and is now neck-and-neck with that of the United States. But, says Paul Maycock, former head of the U.S. solar program, "they'll soon be number one—there's no doubt about it." Maycock left government after the Reagan administration slashed the U.S. solar budget from $153 million to $27 million in a single year. In the wake of a decade of such solar research bloodlettings, the number-one producer of solar cells in the United States, ARCO Solar, was sold in 1990 to the massive German conglomerate Siemens. Maycock estimates that in 1991 nearly 80 percent of Siemens' U.S. production was exported to Germany.

As important as hydrogen may be two decades hence, Germany also understands that the future is now, and has firmly fixed its sights on more immediate targets: emissions, recycling and recapture of materials and energy, corporate marketing, and consumer incentives, as well as international aid. The aggregate effect of these programs is to arm Germany with a wide range of technologies and skills that will, it believes, be in increasing demand as other nations seek ways to protect the environment—a result Germany considers inevitable.

Phasing out ozone-destroying chemicals. In 1989 Germany mandated a ban by 1995 on chlorofluorocarbons (CFCs, sometimes also referred to as Freons, the trade name given them by their inventor and the world's largest producer, Du Pont), the primary destroyer of the stratospheric ozone layer that shields Earth from solar radiation. At the time, the rest of the world had agreed on a global ban, but one not to take effect until the year 2000. Germany, however, opted to ban the chemicals five years earlier, largely because it thought there wasn't time to wait. In June 1990 the nations of the world agreed to push the ban forward to 1996, leaving German manufacturers with a head start in the technologies involved.[26]

Curbing global warming. Germany's commitment to reduce emissions of carbon dioxide, the principal cause of global warming, by 25 percent by the year 2010 will result in one of the world's swiftest and toughest phase-downs, requiring quick response by German industries.[27] But if (most German officials would say when) other nations begin to follow suit, they will find their industries trailing Germany's.

One way of quickly curbing carbon dioxide emissions while increasing productivity is to put energy which is now being wasted to some useful purpose. For example, in most power plants and factories, only about one third of the energy extracted from coal, oil, or gas is actually used. The rest is vented to the air as waste heat. Now, in order to meet its self-imposed goal of reducing CO_2 emissions, the German government is drafting regulations that will require large and medium-sized industries and utilities to market this waste energy. Instead of being vented, it can be used to heat homes and factories (or, by running CFC-free "absorption chillers," cool them), operate paper mills or chemical plants, or even to generate a few more kilowatts in super-efficient power plants. Whatever the use to which the previously wasted heat is put, officials estimate that using waste heat can boost a given power plant's efficiency to roughly 90 percent, cutting air pollution in Germany (already at the world's lowest levels) by 50 percent or more.

Another way to curb carbon dioxide emissions is to develop industrial processes that are inherently cleaner and better. Here again, Germany is in the forefront. Corex, a German company, has developed and built the world's first "cokeless" (and hence less polluting) steel mill. Long considered an essential step in the manufacture of steel, "coking" is the heating of scarce and expensive metallurgical coal in massive airtight ovens. Coke ovens spew a noxious mixture of cancer-causing pollutants that seeps into nearby neighborhoods. They also emit huge quantities of carbon dioxide. Because the Corex process uses ordinary coal to make steel without coking, it both reduces air pollution and allows nations that possess extensive coal reserves, but lack the uncommon metallurgical coal, to get into the steel-making business. By using its pollution control expertise as a lever to open national doors, Corex—and Germany—stand to capture a sizable chunk of the global market in new or rebuilt steel-making plants. The first Corex pilot plant was built in Germany, but a second, full-size facility has already been built in South Africa, and negotiations are underway to build another—in the United States.

The take-back waste program. With landfill space shrinking and trash piles growing, many nations have adopted recycling programs, but none approaches Germany's new "take-back" program, mandated by the Ordinance on the Avoidance of Packaging Waste, enacted June 12, 1991. "The ferocity of the new regulations is extraordinary," said the international business magazine the *Economist.*[28]

Designed to not only reduce trash but increase manufacturing efficiency, the take-back law requires manufacturers and distributors of everything from yogurt to cameras not only to "accept the return" of packages and goods, but then to reuse or recycle them. Explaining the new requirements to a group of U.S. business executives in June 1991, German environmental official Thomas Rummler said they were aimed at achieving "intensified and sustained progress" towards avoiding waste by making reuse and recycling "integral elements of competition." "Those who save costs by offering durable products which can be repeatedly reused," he said, "should enjoy the commercial benefits of the market."[29]

By imposing the burden of dealing with waste on the manufacturers that create it in the first place, the take-back law is intended to reduce—not merely recycle, but eliminate altogether—up to one quarter of the residential waste stream in Germany, thereby saving energy and materials. What waste isn't eliminated is to be recycled, decreasing air pollution and slimming the national economy even further, because products made from recycled materials require up to 95 percent less energy than those composed of virgin matter.

Like most of Germany's environmental laws, the take-back law imposes explicit numeric requirements: by 1995, 90 percent of glass and metal waste and 80 percent of discarded paper, board, plastics, and laminates must be recycled. Incineration, even if used to generate power, has been ruled out as a solid waste disposal method.

Because the take-back law sweeps virtually every form of waste into its ambit, its effects were almost immediate: 400 companies randomly surveyed less than eighteen months after the law took effect on December 1, 1991, said they had completely abandoned the use of polyvinyl packaging, plastic foams, and 117 other types of packaging. Of the 146 companies that had previously used "blister packs," only one still did. One of every four companies was using packaging made at least in part from recycled materials.[30] Companies were running full-page newspaper advertisements touting the recyclability of their products.[31] And for good reason—almost two thirds of Germany's consumers shop for environmentally friendly products.[32] Indeed, public insistence on recycling has become so widespread in Germany that the nation's recyclers were expecting to collect 250,000 more tons of trash than they could handle in 1993. The amount of recycled plastic rocketed from 41,000 tons in 1992 to ten times that amount a year later.[33] In Germany, reported the *Los Angeles Times*, "environmental correctness has come to rival tidiness and punctuality as a national obsession."[34]

The Blue Angel. Reliance on the market has become the hallmark of many German environmental programs, one of the oldest of which is the government's Blue Angel environmental labeling program. Introduced in 1977, the environmental label universally known as the Blue Angel is a symbol owned by the government's environment ministry. The German government describes it as a "market-oriented instrument of government" that relies on "information and motivation, on conviction and the environment conscious thinking and acting of manufacturers and consumers."[35] The ministry licenses the label's use for about 3,500 products selected on a case-by-case basis by the independent nine-member Environmental Label Jury. The label depicts a blue figure with outstretched arms, encircled by the laurel wreath of the United Nations. Inscribed in the border for each product is a brief explanation of the product's qualities—for example, "low-polluting," "low-noise," or "100 percent recycled paper." Although there are imitators in other countries (Green Seal in the United States and the Green Cross of Canada, for example), Germany's remains far and away the most famous and successful environmental labeling program.

The Blue Angel offers its bearers the prospect of winning an edge over competing brands and products in the environmentally conscious German marketplace. The prospect of its award has unleashed a torrent of innovation among manufacturers, and Germany's marketplace boasts a collection of environmentally friendly products, ranging from low-polluting paints to mercury-free batteries.

Surveys show that four out of five German consumers recognize the Blue Angel, and of these, five out of six link it with environmental protection. Public recognition of and enthusiasm for the Blue Angel program have boosted the market share of many products. For example, before water-soluble lacquers were awarded a Blue Angel in 1981, the products commanded a miserly 1-percent market share; today, 40 percent of do-it-yourselfers and 20 percent of professionals buy the less polluting coatings. Similarly, biodegradable chain-saw lubricants, first introduced in 1987, eliminate up to 7,000 tons per year of highly toxic oil and heavy metals otherwise absorbed by forest floors and nearby streams; after receiving a Blue Angel, these oils achieved what the government calls "a dominating position" in the market.

Green aid. Germany's fierce, single-minded focus on environmental protection has enabled it to develop a wide range of products that can be marketed in developed and developing nations alike. To enhance the economic prospects, Germany has mounted a "green aid" program of foreign aid specifically aimed at spurring demand for environmental products and expertise.

If current trends continue, developing nations will account for 40 percent of the world's energy consumption by the year 2010. Germany is betting that much of this new demand will be for the sort of highly efficient, low-pollution technologies that it is fostering. In 1989–90 alone, Germany spent nearly $1 billion in foreign aid devoted to environmental protection, much of it tailored to stimulate demand for German technologies and services. In countries from Kenya to Peru, for example, solar-powered lights, wind-driven pumps, and other nonpolluting sources of energy—many made in Germany, of course—are being introduced at a cost of roughly $140 million.[36]

The cumulative result of these many programs, ranging from green aid to improving scrubbers, is a widening gap between Germany and its industrial competitors, especially the United States. In Europe this gap is not so large because other nations are following suit. Sweden, for example, has adopted a range of pollution taxes, while Austria has embraced a regulatory program even stricter than Germany's. The Dutch are considering a pollution tax while developing sophisticated technologies rivaling those of Germany (one new system, for example, will convert sewage into electricity.)[37]

The gap between the United States and Germany, however, is large and widening, which ought to be cause for alarm within the U.S. industrial community. "Tough standards trigger innovation and upgrading," says Harvard Business School economist Michael Porter, author of *The Competitive Advantage of Nations*, an 855-page multinational study of industrial economies. "Although the United States once clearly led in setting [environmental] standards, that position has been slipping away. Today the United States remains the only industrialized country without a policy on carbon dioxide, and our leadership in setting environmental standards has been lost in many areas."[38]

"Germany has had perhaps the world's tightest regulations in stationary air pollution control," Porter continues, "and German companies appear to hold a wide lead in patenting—and exporting—air pollution [control] and other environmental technologies."

Chapter Two

Japanese Opportunism

> Japanese industry believes that there is an inescapable economic necessity to improve energy efficiency and environmental technologies, which also reduces costs and creates a likely profitable world market. The potential profit in such a market is limitless.
>
> —Takefumi Fukumizu,
> New Energy and Industrial Technology Development Organization, Japan

There had been other steel mills, but none like this one. The typical U.S. steel mill is blackened by soot, streaked with rust, and sagging from too many years of feeding the assembly lines of Detroit and the rest of the rust belt. But here in Japan shrubs lined the manicured lawns spreading like fairways from the plate-glassed, whitewashed buildings. Workers in white smocks and fluorescent-yellow hard hats darted through glass double doors.[1]

This was the Chiba Works of Kawasaki Steel, and in many ways it seemed emblematic of the divergent paths followed by the United States and Japan. At the Chiba Works and in other iron and steel plants in Japan, oil prices and environmental policies had combined to force almost unheard-of reductions in air pollution and oil consumption over a twelve-year period, bringing about new levels of productivity and competitiveness.[2] Between 1970 and 1980, the Japanese steel industry as a whole cut air pollutants by 30 to 80 percent, investing up to one in five dollars in any given year on environmental controls. Since 1980, the industry has invested $11 billion in facilities to protect the environment and another $10 billion in energy-saving measures. As a result, Japan produces steel with 40 percent less energy expenditure on average than the United States and 10 percent less than Germany, its closest competitor. Meanwhile, the U.S. iron and steel industry has continued a steady economic decline, with little substantial reduction in air pollution.

Construction on Chiba began in 1951; it was the first postwar integrated mill (one that starts with raw ore and finishes with rolled steel) to be built in Japan. Expansion continued into the 1960s, then halted until three roughly simultaneous events sent shudders of change through Japanese industry.

The most obvious of these events was the massive economic shock triggered by the first oil embargo, a shock felt by all industrial nations. Equally important, however, was the enactment of two new Japanese laws: a clean air act, similar to the one adopted roughly three years before in the United States, requiring companies to steadily reduce their air pollution; and a program that imposed a tax on industrial air pollution for a special trust fund dedicated to paying medical bills and monthly stipends to several thousand people whose health had been ruined by the air pollution that choked Japanese cities in the 1950s and 1960s.

Though parts of the Chiba Works had been finished barely a decade earlier, these three forces combined to compel its reconstruction. By June of 1977, a "blowing in" ceremony for the new Furnace Number Six (which had a capacity of 10,000 tons per day, enough to make steel for about 20,000 cars) commemorated the first of several major steps in the plant's refurbishment.

At the control room for Number Six, two men seated at a six-foot console occasionally flipped switches as sensors fitted at 800 points in the furnace complex sent data to a computer that made instantaneous judgments and issued operating instructions. The computerized system was, said a company video, "an outstanding example of Japan's technological exports [which] is now being sold to steel works in other countries."

Chiba was also fitted with the world's first industrial-scale selective catalytic reduction system for sharply reducing emissions of air pollution, which technology would be successfully demonstrated at Chiba, then sold to hundreds of utility power plants and industrial facilities throughout the world. Kawasaki's "KM-Cal" process for the multipurpose continuous annealing of sheet steel, also developed at Chiba, was being exported to Europe.

Many of Chiba's improvements, however, were directed not at exports, but at simultaneously improving product quality while increasing efficiency. At the Number Three Steel-making Shop, ladles the size of duplex apartment buildings poured crimson-orange molten iron into Japan's first bottom-blown basic oxygen furnace. By "blowing in" oxygen from the bottom, temperature and reactions can be better controlled, which is, Kawasaki says, "how steel of top quality is obtained." At the same shop, "continuous casting" was added, which eliminates the time- and energy-consuming process of reheating steel after it has been cast; it is, said the company, "part of further development towards complete automation."

Later in the process, slabs of hot metal are run through steel roller pins, flattening twelve-by-twenty-four-foot rectangles into rolls of sheet metal nearly half a mile long. Some are then run through cold mills at eighty miles per hour and flattened into paper-thin sheets. In both mills, computers monitor every step, adjusting speed, pressure, the flow of cooling water, and the other variables that spell the difference between steel of high quality and lesser products. Some of the sheet metal is galvanized to manufacture rustproof cars or enameled tubs. At every step, efficiency is the byword.

Outside, the air shimmered above Chiba's coke ovens, where coal is cooked in airtight containers to drive off contaminants. In most U.S. mills, a fine haze would surround such ovens, billowing over galvanized fences into surrounding neighborhoods; the ovens themselves would be emblazoned with signs warning workers that coke oven emissions cause cancer. At Chiba, however, the air around the ovens was free of this pollution because the coke is quenched, or cooled, with gases instead of water. The heat that in many U.S. mills would be vented into the air in clouds of pollutant-laden steam was recovered at Chiba, then used to generate electricity.

Reconstruction at Chiba, as at most other Japanese mills, is never complete. But in 1985, when officials were sharing these technological revelations with American visitors, they concluded with remarkable statistics: through its reconstruction program, Chiba had slashed air pollution by up to 98 percent and oil consumption by 91 percent.

Standing outside the mill's gleaming headquarters, one amazed visitor asked a plant supervisor where Chiba's managers had learned to do it all. His answer, delivered with a toothy grin and only two words, sent a chill through the visitor: "Republic Steel."

(Already beset by economic woes and $700 million in debt, Republic Steel was acquired in 1984 by LTV, the massive U.S. conglomerate. Barely two years later, LTV filed for bankruptcy, defaulting on $2.6 billion in debt and dumping $2 billion in unfunded pension liabilities on the U.S. government. Most of Republic Steel's old plants were shut down, though one or two were acquired by former employees and continued to operate.)[3]

Chiba's success is mirrored in thousands of power plants and factories throughout Japan. As one Japanese official explained at a 1992 conference in Newport Beach, California, this was not accidental: Japan made a national commitment in the mid-1970s to reduce pollution and increase efficiency through a combined public and private effort.[4]

During the twenty years that followed, the nation spent $60 billion on pollution control, $7.4 billion in 1975 alone. Unlike the United States, where industries avoided the installation of equipment in favor of simply building taller and taller smokestacks, Japanese companies purchased hundreds of devices for eliminating pollution. By 1989, Japan had installed three times as many flue-gas cleaning systems as the rest of the countries of the industrialized world combined. The number of systems for removing sulfur dioxide rose from 323 in 1972 to 1,810 in 1988. Selective catalytic reduction systems for NO_x control jumped from 5 in 1972 to 379 in 1988.[5]

As a result, emissions of sulfur dioxide, the principal cause of acid rain, and oxides of nitrogen, which cause both acid rain and smog, were slashed by about 88 and 80 percent, respectively.[6] By 1989,

Japanese industries emitted less of these pollutants per person than any other industrialized nation—roughly 90 percent less than the United States and one sixth of German levels.[7]

The conventional wisdom is that such massive spending will dampen a nation's economy, yet Japan's continued to grow apace—not despite environmental protections, but because of them. "Contrary to a common belief that implementation of environmental measures will hurt the economy, [such measures] reduced production cost and helped create new markets for energy and environmental technologies," said Takefumi Fukumizu at an international conference on "Clean Air Business Opportunities." The representative to Washington from Japan's New Energy and Industrial Technology Development Organization (NEDO) added that "Japan's business sector has learned not only that environmental protection and economic growth can coexist, but that there is a large potential market for environmental technologies."[8] A study conducted by experts from the Environment Agency of Japan documents that the lesson learned at Kawasaki Steel—that rapid growth and large environmental investment can be complementary—applies throughout the economy.

> The pursuit of strict pollution control measured during the 10-year period between 1965 and 1975 had almost no effect on the macro economy. Investment in pollution control equipment by the private sector during this period totaled 5.3 trillion at 1970 values. . . . [Yet] the real GNP is estimated to be slightly higher when pollution control measures are in effect than when they are not. . . . The economic cooling brought about by a rise in prices from pollution control measures was canceled out by the buying effect of an increased demand for pollution prevention–related facilities.[9]

In an area roughly the size of California but without its rich agricultural lands or domestic sources of fossil fuels, Japan has had to substitute knowledge and ingenuity for resources in order to be competitive. By virtually every measure this tiny nation uses less energy per unit of production—about half as much as the United States—and hence produces less air pollution than any other industrialized nation in history.[10]

This feat is a direct challenge to conventional economic thinking about sources of competitive advantage, which holds that a nation's lack of resources should be a serious handicap because it means other countries can obtain basic commodities at less cost. Strict energy and environmental regulations should, in theory, only exacerbate the problem, diverting capital from more productive investments.

The Japanese experience demonstrates that technology, innovation, and government policy can overcome resource and regulatory constraints. As Michael Porter of the Harvard Business School concludes, "The nations with the most rigorous requirements often lead in exports of affected products" through reengineering of production methods to produce a better product at less cost.[11]

Japanese industry had grown rapidly during the 1960s with the benefit of low oil prices, and the massive conservation programs brought on by the oil crisis of 1973 addressed virtually every aspect of Japanese production, from home refrigerators to steel mills. Conservation ranged from simple acts (the insulation of buildings, for example) to complex and expensive undertakings such as the development of alternative energy technologies. Some results:

- Fuel cells developed or operated by the Japanese generate electricity with zero emissions of sulfur dioxide and levels of oxides of nitrogen below the limit of detection.
- Gas-fired power plants that are among the largest in the world, at a 2,000-megawatt scale, are also among the most efficient—47 percent compared to 38 percent for a typical new U.S. plant. Emissions performance is equally impressive: emissions of oxides of nitrogen at 10 ppm (parts per million), which makes the plant one of the world's cleanest.
- Steel mills built or reconstructed in the mid-1970s were subjected to aggressive energy-saving efforts, reducing the fuel demand of these already highly efficient plants by another one third or more, thus making them the cleanest and most productive in the world.

Turn to the Ministry of Foreign Affairs handbook on Japan's environmental policies and you'll find a proud boast in big, bold type:

"Japan's carbon dioxide emission is about 4 percent of the world's total and one-fifth that of the United States." The figure is so low because Japan uses less energy per person (and per dollar of GDP) than any other advanced nation, thanks to the rapid response of industry to the 1970 oil shock.[12]

"In retrospect," says Genya Chiba, who runs the Science and Technology Agency's Exploratory Research for Advanced Technology (ERATO) program, "the oil crisis was valuable in that it compelled Japan to draw on both technology and a flexible management system. As a result Japan was propelled into the emerging era of conservation and efficiency much faster than the nations that were less threatened. In some cases, within 2 or 3 years industrial oil consumption decreased by 20 to 30 percent."[13]

During the same period in which these improvements were being made in energy efficiency and diversification of supply, and despite enormous economic growth, Japan also made major reductions of air pollution, including the installation of several hundred denitrification units (more than in the rest of the world combined) and the lowest rate of sulfur emissions of any industrialized country.

By the 1990s, Japanese government and industry had developed a menu of technologies and practices which demonstrated beyond question that pollution—even carbon dioxide emissions—could be cut substantially in ways that increased efficiency and lowered costs. These Japanese efforts challenged the conventional wisdom among most American scientists, engineers, and politicians that pollution is the inevitable consequence of industrial productivity. In fact, they suggested—as did German efforts—that just the reverse was true: the path to true productivity is one where the goal is zero pollution and 100-percent efficiency.

These Japanese energy generation and conversion technologies are now likely to be widely deployed if there are to be international efforts to reduce carbon dioxide emissions on a global scale. Economist William Cline of the International Institute for Economics has already warned that, in contrast, the United States could suffer a competitive disadvantage should it fail to follow Japan's example.[14]

"In the past, Japan was criticized because we did not create new technology," says Ikuo Tomita, director for global environmental technology at the Ministry of International Trade and Industry (MITI). But as Tomita lists the innovative, high-technology environmental projects that MITI has begun as part of its thirty-year strategy for cutting greenhouse gas emissions, it's clear that he is determined to put the lie to that criticism. "This program is completely different," he says. "Japan wants to be a leader in global environmental technology, and is well positioned to achieve its goal."[15]

Linking Domestic Environmental Goals to Export Markets

When you have relatively few resources and a small population, economic success demands aggressive pursuit of international trade. Perhaps more than any other nation, the Japanese approach to environmental technology is predicated on this link to export opportunities.

If U.S. standards require steep reductions in emissions from cars, then Japan will make cars that comply. Indeed, to facilitate this process Japan adopted the stricter U.S. standards in the 1970s and maintained them even when standards were weakened in the United States. (To this day, they are referred to in Japan as "the Muskie standards" in honor of Senator Muskie, principal sponsor of the Clean Air Act of 1970.)

The typical Japanese response to international environmental problems has been to see them as part of a larger strategy of designing products to meet the demands of world markets. In these terms, responses to ozone depletion and global warming are much more than environmental policies; they become, in effect, production standards. Adopting consistent domestic standards is essential to helping industry learn what is necessary for success globally. To resist and contest these issues in the fashion typical of American trade associations is to risk the loss of competitive advantage for questionable short-term savings. A senior official of MITI explained, "Japanese industry is participating and cooperating positively in all of these various moves [to promote global environmental protection]. The Keidandren [Federation of Economic Organizations] which is Japan's top business organization has published a global environmental charter. Most major corporations have set up special units for dealing with global environmental issues."[16]

The contrast in corporate philosophy is evident in the difference between U.S. and Japanese corporate responses to global warming. In the United States, electric utilities and other large energy users formed a trade association to lobby against "hasty" action. Indeed, a few utilities mounted a no-holds-barred campaign to stave off the possibility of a governmental response to global warming, including radio and newspaper ads tested in three markets by a bogus organization called the Information Council on the Environment (ICE). Public disclosure in the *New York Times* of ICE's true nature as a sham funded entirely by the coal and electric utility industries led to quick termination of the campaign.[17]

In contrast, the response of the Japanese Federation of Electric Power Companies to growing concern over global warming was to adopt a policy calling for "studies on a number of diverse measures" and committing itself to "contribute at the international level through techniques aimed at improving the efficiency of power production, energy-saving technology, and carbon dioxide reduction technology."[18]

The Federation statement listed a full range of energy "technological breakthroughs" which would be adopted in the future "to conserve energy resources and reduce costs."[19] The two-page list of technological measures ranged from increased use of phosphoric acid fuel cells to measures that chemically separate, collect, solidify, and use carbon dioxide.[20] Although the statement concluded with a brief comment regarding a "lack of scientific understanding," at no point did it oppose, or urge delay in, actions to reduce greenhouse gases.[21]

It is worth noting that environmental politics are much different in Japan than in the United States.[22] There are many local Japanese environmental groups, but few at a national level and none with any true political clout equivalent to that of, for example, the Sierra Club or The Wilderness Society in the United States. Formation of such organizations requires government approval, and contributions to them are not tax deductible. There is also much less opportunity for bringing environmental lawsuits, and global environmental issues are simply not a pressing public concern. Thus, if Japanese business leaders are more concerned about the environment than their counterparts in the United States, it is because they more clearly see its strategic importance, not because they seek to placate citizen or consumer demand.

It is also no coincidence, nor is it a simple matter of altruism, that Japan has been rapidly increasing its environmentally oriented foreign aid.[23] For example, Japan has become a major source of help in Mexico's efforts to improve air quality in Mexico City. The 1993 budget for international environmental energy programs totals over ¥27 billion ($246 million).[24] The benefit to Japan is access to enormous markets for pollution control equipment.

Public/Private Cooperation Japanese Style

The Japanese approach to supporting the development of environmental technology is based on a tradition of deeply interwoven relations between government and industry. Some aspects of the Japanese model are clearly ill-suited for adaptation to the United States, where antitrust laws and a tradition of distrust of government are inconsistent with strong government direction of the economy. However, some key features of the Japanese system are much less "foreign" than is often thought, and the extent of government cooperation with defense-related industries in the United States illustrates that such strategies have been successfully, if selectively, used when national interests were strong enough.

The extraordinary success of Japan's technology policy is the outcome of many different strategies, rather than that of a single model or agency.[25] Careful planning is a consistent theme. Japan first laid out its comprehensive plans for a technical fix for global warming and ozone depletion in 1990, in MITI's "New Earth 21 Action Program for the Twenty-First Century." "New Earth 21" is a vision, as the ministry calls its long-term strategic plans, whose ultimate goal is "to undo the damage done to the earth over the past two centuries, since the Industrial Revolution."[26] But this is no utopian dream: like the "visions" that preceded Japan's conquest of world semiconductor markets, it signals a consensus in industry that it is time to move into a particular area of technology. And, like earlier visions, this one has spurred the formation of new government-industry-university joint research institutes, in this case aimed at transferring existing "clean" technologies to developing countries.[27]

With respect to energy, most technology policy has been overseen by a quintessentially Japanese government-business agency known as NEDO, the New Energy and Industrial Technology Development Organization. NEDO was established by the government of Japan in October of 1980, following the second major oil shock, with a clear mission—the commercial development of new energy technologies which would free Japan of its dependence on imported oil. In 1988, that mission was expanded to include the development of rapidly progressing new industrial technologies in areas such as biotechnology and new materials. In 1990 the mission was expanded once again to explicitly include the development of technologies for protection of the global environment.[28]

Although NEDO is sometimes described as an agency of the Japanese government, it is in fact a quasi-governmental organization; though it was created by and exercises some of the powers of the government, it is not an official part of it. NEDO describes itself as unique because it "*coordinate[s]* the funds, personnel and technological strengths of both the public and private sectors" (emphasis added).[29]

Of NEDO's staff of 846 in 1990, about 250 were assigned to the agency's energy technology efforts.[30] Roughly one third of its energy development employees were drawn from industry, slightly less than one third from government, and slightly more than one third were permanent employees of NEDO.[31] The government employees typically resign their government posts (usually at the Ministry of International Trade and Industry) to serve at NEDO for two to three years, then return to jobs in the government (again, usually at MITI).[32] The industry personnel are from various private companies, such as utility and oil corporations. Corporate employees are assigned to NEDO for two- to three-year periods, during which time their salaries are paid by the government.[33] "Many, many" private corporations assign employees to NEDO, according to one corporate executive.[34] NEDO thus operates with government money and, in part at least, private corporate employees: business contributes people; the government, money. This pattern—government money paired with corporate staffing—is common throughout Japanese projects.[35]

NEDO was assigned responsibility for developing technologies along two complementary paths. The Sunshine Project, according to a senior MITI official, is "aimed at developing technology to discover new and recyclable sources of energy."[36] These include alternative energy sources such as solar, wind, and geothermal power. The Moonlight Project is "aimed at developing technology to achieve greater energy efficiency," as well as the development of nuclear energy.[37] The fiscal year (FY) 1993 budget includes more than ¥34 billion for energy conservation technology, or $314 million.[38] NEDO's purpose is to promote these new technologies "to a level at which private industries can take over and commercialize them for themselves."[39]

With the expansion of its mission in 1988 to include industrial technologies, NEDO began to focus on areas which were unrelated or related only indirectly to energy. NEDO describes these as "technologies which have high development risks and require long lead times for basic development." Examples include a manganese-nodule mining system and a system for intravenous laser surgery. One specific technology related to global warming that NEDO has sought to develop is methods for the fixation and utilization of carbon dioxide.[40]

In fields related to the development of new energy, NEDO's Sunshine Project has targeted a wide range of technologies, including solar cells with high conversion rates and high efficiencies, ways of using these solar cell systems for power generation which are commercially practicable (that is, able to integrate with the utility grid), and a large-scale wind power system. The "New Sunshine" program included in the FY 1993 budget is to receive ¥53.9 billion, or $490 million, an increase of more than 7 percent from FY 1992.[41]

In pursuit of "Moonlight" increases in efficiency, NEDO made priorities of coal liquefaction and gasification technologies, fuel cells, high-performance heat pumps, and electricity storage technology.[42]

Japan also retains a strong commitment to nuclear power and has maintained an enviable operating record at its forty-one reactors. Nuclear power is projected to be an important factor in limiting the buildup of carbon dioxide, the most important greenhouse gas. In FY 1993, nuclear power research and promotion is budgeted for more than ¥950 billion, or $8.6 billion.[43]

MITI announced in 1990 a series of new initiatives designed to improve old technologies, develop new ones, and sell all of them to developed and developing nations alike. All of these fit under the "New Earth 21" umbrella.

The Research Agency of Innovative Technology for the Earth (RITE)[44] is a new agency that may assume responsibility for some NEDO projects. RITE is one of several new Japanese quasi-governmental organizations created in 1990. Together, these organizations provide Japan with the opportunity to seize unchallenged global industrial leadership in a host of areas.

RITE is intended to develop the environmental technologies "New Earth 21" says will be needed early next century. Instead of searching for ways to improve energy efficiency (with lowered carbon dioxide emissions as a lucky byproduct), RITE is aiming toward schemes for stripping carbon dioxide out of industrial emissions and recycling the carbon.[45]

As a "public foundation," RITE is allowed to accept government funds for the conduct of projects.[46] RITE was funded with ¥8 billion (about $50 million at 1990 exchange rates) in donations from regional and municipal governments and private businesses. Corporate funding was provided by two umbrella business organizations, the Federation of Economic Organizations (the Keidandren) and the Kansai Economic Federation. They, in turn, allocated shares of support to various industry segments. Roughly ¥5 billion ($30 million) was donated by business and the remainder by governments.

Industry has responded enthusiastically. In the half-year after the institute was set up, $45 million came in from Japanese industry, adding to the $80 million in seed money provided by MITI. Every sector is represented: the huge electric utilities, engineering companies, car manufacturers, shipbuilders, electronics companies, steel manufacturers, and even clothing makers. All have their own environmental technology programs, too—Tokyo Electric Power, for example, is searching for ways to extract carbon dioxide from smoke by chemical absorption—but they see involvement with RITE as essential to keep abreast of long-term technological change and the latest in government thinking.[47]

Thanks to industry's donations, RITE is now able to spend about $28 million a year to support work by about 200 researchers. Half of the money goes toward the two big carbon dioxide–fixation projects and the rest is divided among the remaining big projects, subsidies to private companies for the development of environmental technology, and grants for basic research. For the time being, all of the work is going on at the laboratories of the companies, universities, and MITI institutes that have joined RITE projects. But next year this will change when RITE's own laboratory is completed in the new Kansai Science City under construction near Osaka.[48]

RITE's research efforts are divided into two components. One is concerned with the identification and development of a substitute refrigerant as a replacement for ozone-destroying chlorofluorocarbons and hydro-chlorofluorocarbons. The second deals with three somewhat unrelated projects: identification and development of methods for biological or chemical fixation of carbon dioxide, development of biodegradable plastics, and development of methods of synthesizing chemicals biologically.

International participation in all RITE projects is welcome—a highly unusual arrangement for a Japanese institution—but at a price: intellectual property rights on the fruits of research must be shared with RITE. Last year, applications for grants came in from the United Kingdom, the United States, Canada, Australia, and the Netherlands, says Hidefusa Miyama, director of the Research Planning Department of RITE. Already an Italian company has joined one large project on microbial generation of hydrogen, and a group at the British Agriculture and Food Research Council has won funding to study the uptake of methane by agricultural soils.[49]

RITE does not fund its projects outright; rather it manages research projects with members drawn from governments, corporations, and universities. RITE utilizes funds made available by government agencies such as NEDO for these tasks. Typically, a task group is formed to undertake research in a given area. Individual researchers are assigned to the RITE project by their employers for a period of roughly two years. The researchers continue to work in their own company or in university labs and receive salaries and benefits from their employers. They are, however, entitled to assume the title of "researcher" for RITE. RITE's research budget for this kind of activity in 1991 was approximately ¥3.5 billion ($2 million).

In addition, RITE makes matching grants to support specific research projects at corporate and other laboratories, universities, and nonprofit organizations. RITE can provide up to 50 percent of the funding for such projects. Its approximate budget for this task in 1991 was ¥1.7 billion ($1.1 million). Finally, RITE provides small research grants to "incubate" ideas. About ¥60 million ($550,000) was allocated for this function in 1991.

The new technology being developed at RITE won't be ready for twenty years; until then, Japan sees the transfer of its existing energy-efficient technology to the developing world as the best way to tackle global warming. The logic is simple: most industrial countries can now hold their carbon dioxide emissions at current levels or start to reduce them, says Ikuo Tomita of MITI, but industrializing countries going for quick economic growth to support growing populations will pump out more greenhouse gases every year. He cites World Bank figures that show that developing countries will produce 44 percent of total carbon emissions in 2050, up from 20 percent as of 1992.[50] The trick is to make efficient technology available at a price developing countries can afford, while demonstrating that environmental protection is consistent with short-term economic goals.

Japan's answer to the challenge is the International Center for Environmental Technology Transfer (ICETT), which won MITI backing last year. Its objectives are ambitious: Tomita says it and an associated energy center are going to try "to train 10,000 people over the next ten years" in energy conservation, pollution control technology, and environmental protection regulations. Most participants will come from developing countries, and as with other Japanese overseas programs, many will end up working for Japanese companies when they get back home. The benefits to Japan include entry to new markets; after they have been trained, participants' first choice of technology is likely to carry the "Made in Japan" label.[51]

If the creation of ICETT and RITE were not enough to worry those concerned about Japan's growing technological dominance, there's yet another unsettling prospect—a new United Nations Environment Program International Environmental Technology Center (UNEP/IETC), first proposed by Toshiki Kaifu, then Japan's prime minister, at the Houston summit of advanced nations in July of 1990. Like RITE and ICETT, it will be built in Kansai Science City and should be up and running by the end of 1992. This added boost for Japan's technology transfer activities caused George Bush's science adviser, D. Allan Bromley, to voice concern that the decision to put the United Nations institute in Japan may mean U.S. industry has missed the environmental technology boat.[52]

Japanese Profits From American Invention

Another striking contrast between the Japanese and American approaches to technology has been the inability of the United States to move technology from basic research to the marketplace. This phenomenon has been repeated many times over. The classic example is the VCR, which was originally conceived and developed in the United States, and demonstrated in early models, but abandoned as a commercial product when no private U.S. company proved willing to commit the capital and the production engineering skills necessary to achieve commercial success. Japanese companies, with their longer financial horizons, access to lower-cost capital, and emphasis on expanding markets, picked up the technology at minimal cost and now dominate the market.[53]

This experience is even more troubling when the technologies have been developed with U.S. taxpayer support, sometimes over a long period and enormous subsidy. Access to the so-called Cool Water technology for gasifying coal and organic material (see Chapter 7), the license to Englehard's fuel cell program, the lead in manufacture of solar cells—all were heavily subsidized by U.S. taxpayers but have now been partly or largely added to the Japanese industrial arsenal and lost to the United States. The dramatic increase in the value of the yen (which appreciated more than 15 percent against the dollar in the first six months in 1993) will only exacerbate this trend, as American technology becomes an even better bargain for Japanese companies.

Japanese companies are encouraged to develop and commercialize technologies through a wide range of incentives. There are three principal corporate tax incentives for technology development. One stipulates that 25 percent of any year-to-year increase in research-and-development expenditures over the previous year is a tax credit, up to a limit of 10 percent of total corporate tax. In 1978, manufacturing companies realized a ¥15-billion (about $50-million) benefit through this credit. The second is accelerated depreciation on research-and-development facilities and hardware, which can often mean a 60 percent write-off of the original purchase price in the first year. These provisions are aimed at conserving the cash flow of high-technology businesses.[54]

The government, in cooperation with other institutions, also directly funds research and development. Such grants can take several forms. Most common are matching grants in designated research-and-development areas, given either to companies or to associations formed for that purpose, often groups of small companies that could not finance new technologies on their own.[55]

Recently Japan has redoubled its efforts to build a commanding lead in the development of new, clean technologies. For example, for commercial enterprises which install photovoltaic (solar electric) systems, 7 percent of the PV system cost is deducted from taxable income. Loans are provided to the company with interest rates as low as 4.1 percent. In 1992 the government set up a new institution to finance PV installations in public facilities (schools, public halls) for two thirds of the total cost.[56]

Under a "special taxation system for certain equipment responding to changes in energy and environmental circumstances," Japan is providing tax incentives for the installation of an enormous range of technologies:

- heat pumps, floor heating, PV solar systems, and solar thermal water heaters in houses;
- cogeneration systems, waste heat recovery boilers, and other energy conservation and load-leveling equipment in businesses;

- low-polluting internal combustion engines in industry generally;
- district heating and cooling systems, fuel cell and combined-cycle turbine systems in power generation facilities.

Under this tax provision, companies installing designated equipment can take a one-time tax credit of up to 20 percent, or a special depreciation allowance in addition to normal depreciation. In addition, companies installing designated equipment get a three-year 17-percent reduction in their municipal property valuation for that equipment.

Japan is also promoting the construction of PV, wind, hydroelectric, and geothermal power generation facilities under a separate "local energy utilization promotion taxation" system. Electric and methanol vehicles are being encouraged through incentives in a motor vehicle tax, a mini-sized vehicle tax, and a motor vehicle acquisition tax. A "model shift promotion taxation" system provides incentives to freight forwarders to purchase containers for rail transport and trucks suitable for piggyback rail transportation (facilitating the use of trains rather than less efficient trucks). By way of disincentives, Japan's gasoline tax is about $1.60 per gallon, compared to about $0.29 in the United States.[57] Japan is also implementing the world's most stringent air pollution standards for diesel trucks.

For the most part, Japan lags behind the United States in promoting conservation programs by its electric utilities. MITI has permitted Japan's electric utilities to adopt an electricity rate system that encourages customers to shift their consumption to nighttime periods, when demand is much less, and therefore reduce the need for new power plants.[58] Electricity consumers that install conservation and load-leveling equipment used at night pay lower electricity rates. For example, over half a million customers have signed up for Tokyo Electric Power Company's thermal-storage electric water heaters with nighttime-only service. In addition, Japan's electric utilities are beginning to set up centralized load-control systems that permit them to directly control electricity flow to equipment in homes and offices, permitting companies to suspend or reduce power supply to designated appliances during peak times.[59]

In short, Japan is responding to the new environmental realities in a single-minded, organized fashion. It is yet to be seen how the United States will react.

The arrival of the Clinton/Gore administration has given some basis for hope that the situation in the United States will improve. A new technology rhetoric linking environmental goals with job creation through strategies such as a "green car" initiative is particularly encouraging.[60] On the other hand, the near total defeat of President Clinton's energy tax proposal indicates that change will not come easily. In any case, Japan is not likely to be easily overtaken, nor are the major European powers. As John Zysman, a professor of political science at the University of California at Berkeley and the author of several books on manufacturing and competitiveness, recently observed, "The conviction is widespread in Japan that it will be the dominant technological power by the end of the century; indeed, the view of some is that the transition has already taken place. In any case, purposive political strategies of economic and industrial development are critically intermingled with Japan's surge to economic power.[61]

The challenge, then, is for the United States to fully mesh its environmental and economic policies and to recognize that increasingly superiority in environmental performance is an element of the definition of competitive advantage. In Japan, this advantage has been won by that nation's unique system of public/private partnership in support of identified national priorities. The U.S. response will have to differ in critical ways that reflect the enormous differences in our culture, political system, and laws; however, the United States retains many advantages due to its size, natural resources, and creativity.

The American system, with its emphasis on individual freedom, continues to be the world's best in terms of invention and innovation. The Japanese and German systems both tend to produce success through incremental change, while the United States focuses on breakthroughs, hoping to leapfrog over its competitors. The strength of the German and Japanese approaches is that they can successfully bring inventions to market, albeit sometimes with a sacrifice of personal creativity. America, too, was once a master of bringing inventions to market—witness products ranging from the Model T to the Xerox machine. The nation's challenge is to relearn those old skills, while building on "Yankee" ingenuity. This is a race that the United States can win, if it wants to.

Chapter Three

The World Market and American Decline

In the future, access to international markets will depend on who has the most environmentally sound technologies. If U.S. companies don't move aggressively, we will see the same conflict in environmental technology that we see today between GM and Honda.

—TSUKASA SAKAI, JGC Corporation

As the world makes the transition from one technological and environmental era to another, it is increasingly difficult to say whether customers—industries or consumers—buy "green" products ranging from turbines to tissues because they are better for the environment, or because they are simply better. It is clear however, that the environmental market is inexorably expanding to encompass all products for all purposes.

The examples of Germany and Japan illustrate the manner in which two of the world's industrial giants are beginning to take advantage of the new economic and environmental realities. The potentials for other economies, especially that of the United States, are enormous, as is the scope of the global market and the range of industries that will be affected.

Even if the environmental market were to remain limited, as it has been since the 1960s, to add-on pollution control devices like catalytic converters to reduce tailpipe air pollution and wastewater treatment plants to abate water pollution, it would exceed $100 billion a year in the United States alone. Globally, it would be much larger, perhaps $200 to $300 billion annually, because demands for catalytic converters and other add-on controls are rising sharply.

If the world remains content with power plants that waste two of every three pounds of coal, cars that dissipate four of every five gallons of gasoline, and containers that require a hundred units of energy to make instead of one, the market will continue to be limited to add-on controls. If, on the other hand, external imperatives such as consumer demand, population growth, threats to the environment, and an increasing scarcity of virgin raw materials relentlessly push the world in the direction of more efficient, less polluting light bulbs, cars, power plants, and other devices—and they will—then the market extends, potentially, to every good and every service, from cosmetics to garbage collection. It is this larger market that the Germans and especially the Japanese have in mind when they consider the profit potential of "environmental" goods and services. It is a market in which virtually all products will possess an environmental component, and consumers will consider products on that basis in much the same way that today's buyers take into account price and quality.

The Environmental Industry

Although the outlines of this larger market are just beginning to emerge, the smaller one is already reasonably well defined, although partly because there is no single definition of "the environmental industry," there is no accord on exactly how much is spent in this market each year.

The U.S. Environmental Protection Agency (EPA) terms the "environmental industry . . . a highly diverse collection of businesses that provide about $100 billion a year in goods and services used to improve the environment."[1] One commentator explained:

> In fact, if environmental protection were a corporation, it would rank far higher than the top firm listed in the Fortune 500. In 1992, expenditures for environmental protection totaled $170 billion. In con-

trast, in 1991, General Motors, the largest U.S. industrial corporation, had sales of $124 billion; the number-two U.S. industrial corporation, Exxon, had sales of $103 billion; and the third-ranked corporation, Ford Motor Company, had sales of $89 billion. . . . In 1992, environmental protection spending created about 4 million jobs [which is] about 3 percent of total 1992 employment.[2]

EPA's definition is narrower than some, because it excludes spending that falls outside the programs within its jurisdiction. Spending for the disposal of solid waste (say, in landfills) is an "environmental" cost because the agency regulates dumps, but since it doesn't regulate recycling, those sums are excluded from EPA totals.[3]

Other estimates are provided by investors. One firm, EnviroQuest, divided the industry into twelve segments and estimated annual revenues for each. It concluded that 1989 revenues totaled $118 billion.[4] The consulting firm Management Information Services, Inc. (MIS), of Washington, D.C., utilizes yet another estimate of annual spending for what it calls the "pollution abatement and control industry," concluding that 1989 spending was $105.6 billion, representing roughly 960,000 jobs.[5]

Despite the differences in methodology and scope, these various estimates all depict the U.S. environmental industry as one which, in the words of MIS, is "an important driving force in the U.S. economy."[6]

Estimates of the size of the global environmental market also vary, with most analysts placing the industry at about $200 billion per year currently, with a growth rate likely to bring it as high as $600 billion by the year 2000.[7] Much of this growth is being driven by demands in developing nations from Asia to South America, where environmental markets are following similar patterns of rapid growth brought on by prolonged inattention to environmental degradation. The *Economist*, terming the potential market for environmental services in Southeast Asia "vast," noted:

Until recently the region's governments have not cared a jot about pollution. Levels of particulate matter (such as dust and grit) in the air in Bangkok, Jakarta and China's big cities exceed the World Health Organization's standard on 100 days or more a year (compared with ten days in most developed countries). Sulphur-dioxide pollution is among the world's worst. Thailand's Chao Phraya river is dead. Jakarta has so over-used its ground-water that salt water is seeping into wells up to ten kilometers inland. In Hong Kong (which means "fragrant harbor"), two million tons of untreated sewage, chemicals and toxic metals are dumped into Victoria Harbor every day.[8]

Now, however, things are changing as Asian governments are belatedly spending lavishly to restore their environments. South Korea has announced plans to spend $12 billion, and Taiwan $10 billion, on cleaning up. Hong Kong plans to build a $2.5-billion sewage treatment plant, and the World Bank intends to lend $4.3 billion over the next four years to help the poorer countries in the region repair their environments.[9]

Singapore has already grown rich by selling services to its larger but less sophisticated neighbors. One niche it has high hopes for is the business of helping other countries in the region clean up after their rapid but filthy industrial revolutions.[10] To develop expertise as well as to repair its own environment, Singapore has already taken the lead in several environmental initiatives. These include a ban on nonpharmaceutical aerosols and polystyrene products containing chlorofluorocarbons, as well as the introduction of unleaded gasoline, accompanied by new emissions limits on motor vehicles.[11] In 1992 the Ministry of Environment allocated $609.3 million not only to improve Singapore's own environmental health, sewage treatment, and solid waste disposal, but also to develop the nation as the regional center and transit point for environmental products and services.[12] Singapore Environmental Management and Engineering has also already won contracts to deal with sewage and waste management in parts of Malaysia and Indonesia. It is also training Indonesia's Ministry of Energy to do environmental impact assessments.

While this pattern of spending growth is occurring most significantly in Asia, it extends throughout the developing world. Mexico, for example, plans to spend upwards of $400 million in the United States–Mexico border region alone by the mid-1990s.

Limitless Profits

Increasingly, however, the phrase "environmental technology" is being associated with products whose functions are not directly related to environmental protection. The surge in "green" consumer products, ranging from detergents to body stockings, is obvious. Less apparent but more important is the rising number of industrial and commercial products: turbines, solvents, lights and lighting controls, industrial motors, heating and air conditioning systems, engines for cars, trucks, and buses, and a virtually unlimited list of other goods.

Two of the largest markets are for the machines that in the aggregate consume the vast majority of the world's energy and generate most of its pollution as well: motor vehicles and electric power plants.

For a long period, the United States and Japan stood alone as the only two nations with aggressive motor vehicle pollution control programs. But the 1980s saw the rest of the world beginning to move toward U.S. standards. That trend has accelerated to the point that 4 of every 5 cars made is now equipped with pollution controls. Nations which lagged behind the United States in 1980 are now seriously discussing programs that would be more stringent than that of the United States.

Initially, antipollution requirements were satisfied through the addition of a few devices such as catalytic converters. However, as standards have become more stringent, the control systems have become the car itself: low-emitting engines, low-rolling-resistance tires, aerodynamically superior designs, lighter-weight body parts, on-board computers, and dozens of other components now constitute the pollution control "system."

Recently, the trend towards "greener" cars has expanded. In California, officials have begun to view the vehicle and the fuel as a single entity. Cleaner, "environmental" gasoline and diesel fuels have come onto the market, while new entrants—natural gas, methanol, ethanol, and electricity—have begun to vie for customers. Simultaneously, Germany has initiated a take-back program that requires automotive components to be recyclable. Both the California air pollution program and the German take-back requirements are spreading to other governments, forcing a market redefinition of what constitutes a green car, in the new sense of the word. No longer is it sufficient to produce smaller quantities of noxious air pollutants; now a car must produce those pollutants only in vanishingly small amounts. It must also be recyclable. Its air conditioner must be filled with "ozone-safe" refrigerants. And soon probably, it will be required to minimize emissions of pollutants that cause global warming.

A parallel though less obvious trend is overtaking the world's electric utility industry. The United States was the first to enact stringent controls on power plants and other industries, followed by Japan. But in the 1980s, a succession of European nations followed suit, enacting controls that were vastly more stringent than those in the United States, excepting California. Germany, for example, required all of its power plants to reduce pollution by 90 percent. As this was happening, California began to forge new industrial regulations as companions to its tightened motor vehicle program. Germany began requiring that waste from power plant pollution control systems be recycled; Sweden adopted a tax on carbon dioxide emissions.

No longer is it sufficient for a power plant to reduce emissions of a handful of air pollutants. Now it must eliminate 90 percent of a wide range of toxins, and seek to go even further. Its wastes must be recycled, and the little pollution that remains is taxed.

In the newly industrialized nations, the demand for electricity—and with it the need for new equipment—is rising sharply as productivity and standards of living climb. In the United States, Europe, Japan, and other previously industrialized nations, the demand for equipment is spurred by the need to replace aging machinery installed in the interval between the conclusion of World War II and the beginning of the 1970s.

The U.S. government's Agency for International Development (AID) estimates that the global market for major energy technology over the next twenty years will be about $2.1 trillion. Over $900 billion will be for power generation, transmission, and distribution equipment and services; $955 billion for oil and gas exploration and development; and $250 billion for energy conservation.[13] Here again, the market for electric power equipment in developing nations is enormous, and growing so explosively that one executive, John B. Wing, partner in Wing-Merrill Group Ltd. of Aspen, Colorado, has termed it "the biggest opportunity any of us will see in our business lives."[14]

Worldwide energy use is projected to shoot up 40 percent, from its 1988 level of 327 quads to over 450 quads by 2010, with the bulk of the increase in the developing countries of Asia, Africa, and Latin America, as well as in Eastern Europe. (A "quad," a quadrillion British thermal units, is the measure often employed by analysts in discussing exceedingly large amounts of energy. The United States, the world's largest energy user, consumes roughly 80 quads per year, while a single quad is approximately equal to the annual energy consumption of Austria.) These countries will have energy-use growth rates of between 2.0 and 2.9 percent per year, nearly double those of the industrialized nations. In fact, by 2025, energy use in developing and Eastern European countries will likely surpass that in the developed OECD countries.[15]

A survey by McGraw-Hill Inc.'s Independent Power Report found potential for 290,000 megawatts of development (a standard coal-fired plant generates about 600 megawatts) in forty-two countries, led by Pakistan, India, China, Argentina, Indonesia, and Thailand. Another tabulation, by *Electricity Journal* magazine in 1992, found 453 projects being developed worldwide to build and operate entire power plants and transmission systems. Little wonder that John Wing rhapsodically exclaimed that "electrifying the world is one wonderful opportunity." Wing's firm, backed by investment capital from utilities, recently signed agreements to build and operate three power plants in China.[16]

China, like many other developing nations, needs electricity but is short on capital, forcing it to open its doors to foreign ownership of basic industry and thus affording U.S. firms like Wing's the opportunity to profit not only from the construction of power plants but from their operation as well.[17] The utility business, once limited to the geographic borders of a specific nation or state, is now able to leapfrog such boundaries. Already, utility companies from Japan to Virginia are poised to become multinational corporations, as did auto, steel, and chemical firms before them. United States utilities have historically been either reluctant or precluded from entering this market, but due to 1992 changes in federal law, "U.S. utilities can now join the chase," in the words of the *Washington Post*.[18]

If the United States were able to retain its historic share of the electric power market, its firms could expect to claim $226 billion of the projected sales. Unfortunately, the AID study concludes that in recent years U.S. trade competitiveness has declined drastically. According to one AID official, "For the Rotating Electric Plant, the U.S. share dropped from 20 percent of world exports in 1982 to about 10 percent in 1987. Meanwhile, its share of world imports of this equipment rose from less than 7 percent to over 14 percent. This pattern of worsening competitiveness is repeated in most energy sectors."[19]

Clearly some of this decline is due to the environmental inferiority of much U.S. equipment, though how much is difficult to say. All of the major industrialized nations have adopted air pollution control laws, many of which explicitly require that the power plant technology selected be the "best available" from an environmental perspective. Many U.S. systems simply fail to meet this test. Moreover, some nations (for example, Germany) prohibit the dumping of waste, even when it is the byproduct of a pollution control system. Here again, U.S. systems don't measure up, largely because U.S. laws allow scrubber sludge and other pollution byproducts to be discarded. Finally, as some nations (for example, Sweden) enact taxes on either pollution or energy, utilities are searching for machinery that is not only the cleanest available but the most efficient. Here again, with the notable exception of General Electric's combined-cycle systems, U.S. equipment falls short of the Japanese and European competition.

If there were some help from the U.S. government on the model of the "green aid" programs of Germany and Japan, American firms might be able to make inroads in these markets. However, the United States significantly lags behind its major international trade competitors in the public resources committed to trade promotion. Among eight major trading nations—Belgium, Canada, France, Italy, Sweden, the United Kingdom, West Germany, and the United States—the United States ranks last or next to last in export-promotion indicators such as spending per capita, spending as a fraction of GNP, spending as a fraction of total spending, and spending as a fraction of industrial exports.[20]

Still, even if the opportunity for exports were eliminated altogether, the market within the United States itself is substantial because of the need to replace power plants so old that they are increasingly inefficient. After peaking in the early 1970s, fossil-fuel power plant capacity additions slowed dramatically. In 1973, for example, 23,000 megawatts (MW) were added to the nation's electric power grid; in 1982, only 7,000 megawatts were added. A consequence of this slowdown has been a steady rise in the age of fossil fuel–fired power plants. In a 1987 study, the U.S. Department of Energy concluded that

> America's fleet of existing fossil fuel power plants is aging rapidly. At the same time, growth in the nation's demand for electricity is projected to increase, surpassing the existing, committed generation capacity within the next 10 years and continuing steadily upward. These two inexorable trends—aging power plants and increasing demand for power—will converge in the mid-1990s, creating a critical period for the nation's utilities and their customers.[21]

A similar situation prevails in Western Europe, where, as in the United States, new power plants are needed to replace old ones that are simply worn out. Demand for replacement capacity in the European Economic Community (EEC) will total 62 GW (gigawatts) in the period 1992 to 2000 and 75 GW between 2001 and 2010, almost as much as the most optimistic forecast of new capacity demand.[22] (The watt, of course, is the basic measure of electricity; 75 watts powers a light bulb. A kilowatt is 1,000 watts, a megawatt is 1 million, and a gigawatt is 1 billion. About 4.5 kilowatts will run the average U.S. home, while a gigawatt is roughly enough electricity for all the homes in, say, Portland, Maine; Knoxville, Tennessee; or Boise, Idaho.)

The European Commission expects new capacity in Western Europe to rise only by between 30 and 190 MW in the twenty years to 2020, averaging 0.3 percent to 1.8 percent growth per year. In contrast, Eastern European capacity is expected to increase by about 32 GW in the period to 2000, a growth rate of about 2 percent per year.[23] There is no assurance that U.S. manufacturers will necessarily claim even the American market for new and replacement equipment. Federal and state laws that historically shielded U.S. utilities from competition have recently been repealed or altered, opening the door to foreign manufacturers, as the Doswell plant illustrates. Europeans are certainly looking to their own companies to satisfy their demand, making it unlikely that American firms will claim much, if any, of this market.

As U.S. and European power plants age, utilities are confronted with hard choices. On the one hand, they can undertake conservation and other programs to reduce electricity consumption. But unless regulators can be persuaded to change the rules in order to allow higher profit margins on declining sales, utility income will drop along with sales. On the other hand, utilities can spend money on the power plants themselves. Their options range from replacing a facility entirely to "life extension," a relatively low cost refurbishment that allows a plant to limp along for another few years. The middle course that many U.S. utilities are opting for is "repowering," or replacement of an aging plant's furnace and boiler with new, cleaner-burning and higher-efficiency technology.[24]

The U.S. Department of Energy has been one of the most enthusiastic advocates of repowering, saying that "a new array of advanced pollution control devices (for retrofitting existing plants) and repowering technologies could be widely deployed beginning in the 1990s. The result could be lower electricity costs for consumers and a significant and permanent reduction in coal-based emission levels."[25]

Thus, all of the options available to utilities result in the selection of a green technology, and many of these technologies (to be further described in later chapters) once were American but now belong wholly or substantially to Germany, Japan, or other nations. The future of electricity generation will belong to these technologies because they can produce power with zero or near-zero pollution while converting up to 90 percent of their fuels into useful energy.

Consumer Products and Public Opinion

The surge in green products that is just beginning in electric power plants and motor vehicles has already begun in earnest with consumer goods. Demand for green consumer products exploded in the late 1980s and early 1990s, propelling companies like Ben &

Jerry's and The Body Shop into the forefront of public consciousness and corporate profitability.

There was a burst of environmental marketing in the late 1980s that subsided somewhat as environmental groups and state governments revealed many corporate claims to be false "green scams." Having matured, and sustained by new labeling and government oversight programs, green marketing underwent a resurgence in 1992, when the green consumer-products market totaled $110.1 billion. It grew 10.4 percent, to $121.5 billion in 1993, according to Carl Frankel, publisher of *Green Market Alert*. Jeanne Meyer, manager of environmental affairs at Lever Bros., explained that "consumers appreciate when companies take real steps to reduce packaging. . . . People may be feeling that environmental marketing is passé, but that doesn't reflect a lack of activity. If anything, companies have stepped up their activity."[26]

As companies have increased the "greenness" of both their products and their claims, outsiders have stepped in to assure at least a modicum of honesty. Germany's Blue Angel environmental labeling program has been joined by the European Community's "eco label," which sports a cheerful flower logo with an *E* for a pistil. In the United States, there is no government program, but two private firms, Green Seal and Scientific Certification Systems, issue their own stamps of approval. Japan and Canada have also launched government programs.

Such labeling programs are evidence of one of the realities propelling industry into the new era, namely consumers themselves. As customers have brought the power of their purses to bear, companies that make and sell retail goods have had no choice but to comply with public sentiment. The simple truth is that whether customers are right or wrong in placing a premium on environmental protection, their beliefs—and buying habits—have become a market reality.

Despite economic crises, oil embargoes and even wars, public commitment to environmental protection has not only endured but deepened. In Houston, Texas, where Steven Klineberg of Rice University has been polling voters since the late 1970s, nearly two of every three voters believe that humanity is approaching the limit of

Earth's resources (the number of people expressing this belief jumped from 56 percent to 63 percent between 1991 and 1992). In early 1992, a Klineberg poll found support for environmental protection spread roughly equally across all income levels, with "people who live in low- to moderate-income areas are just as likely to be concerned about the environment as people in the affluent neighborhoods," according to the *Houston Post*.[27]

These results crop up time and again. Respondents in another poll said they were willing to pay more for environmentally sound products: 47 percent said they were already buying energy-efficient and environmentally sound products, while another 36 percent said they would like to buy them but had difficulty finding them.[28]

Eighty-seven percent of respondents in a 1991 Golin/Harris poll said they would boycott a company that is "careless" about the environment. Rich Jernstedt, president of Golin/Harris Communications, viewed the poll as "sending a signal to government and business that voters are willing to make sacrifices to improve conditions for themselves and future generations." In his view, Jernstedt said, "industry and government must address these key issues and provide leadership in moving environmental measures forward." [29] He added that "environmental consciousness already has had a profound effect on product marketing and promotion."[30]

These results barely scratch the surface of a massive and compelling body of polling data demonstrating a commitment to environmental protection so deep and abiding that it is, in and of itself, a market reality and a competitive opportunity. One of those to whom such polling data comes as no surprise is Willie Parker Goodwin.

Goodwin has what many would consider the perfect job: for twelve weeks every two years, he hop-scotches across the South Pacific from Tahiti to Fiji, staying at $500-a-day beach resorts and dining at posh restaurants. But it's all in a day's work for Goodwin, for the thin, graying North Carolinian is the author of *Frommer's Guide to the South Pacific*. The winter of 1992–93, however, was not quite another day at the beach.

Waiting in the Raratonga airport for his flight to Fiji, Goodwin was stirred from his reveries by an announcement that the plane was being diverted to Aukland to avoid a tropical cyclone. No problem, he just lost a day. But three weeks later another cyclone, Kina, began making its way up west of Fiji. Simultaneously, a third, Hina, began building near Australia. By the end of Goodwin's twelve-week stay, cyclones had hit the Solomons, New Caledonia, Vanuatu, Fiji (three times), Tonga, and Western Samoa. Torrential rains had left hillsides muddy and rutted, and tropical fields were poisoned with saltwater, leaving them a barren brown. Before 1983, Tahiti hadn't been hit by a cyclone in seventy years. But it was struck by three that year, and has been hit regularly since then.

Stronger, more frequent storms are a global phenomenon, and the pain is acute in the wood-paneled offices of the world's insurance companies. From 1966 to 1987, the insurance industry paid no claims over $1 billion. But since then there have been fifteen catastrophes with losses higher than $1 billion. Four storms in Europe in 1990 cost $10.4 billion, while Hurricane Andrew's 1992 toll in Florida amounted to $20 billion.[31] Not surprisingly, Swiss Re, one of the industry's large reinsurers, advises that "it would be prudent for the industry to act as if the [global warming] theory were correct."[32]

This conclusion is one that the voters and consumers of the world, as well as many governments, have already reached. They have concluded that the evidence that some pollutants destroy stratospheric ozone, others cause global warming, and others still cause death and injury is compelling, and that there is no reason to believe that the body of evidence will not grow with time, or that it will point in any direction other than toward the necessity for increasingly stringent pollution controls. Whether the public and these governments are right or wrong is irrelevant from a business perspective. Their attitudes and policies are a market reality that has already reconfigured global business, creating a new era—one in which the United States can be a winner or, absent some change from its present course, a loser.[33]

If the United States were losing the race to capture the global monopoly on plastic whistles, few would care. But at stake here isn't a niche or even a large segment, it is virtually the entire global market—everything from yogurt cartons to tennis shoes. As a senior official of AT & T told the Senate,

> We are talking about restructuring the technological basis of our entire economy to make it sustainable over the long haul. We are not talking about devising better scrubbers; we are talking about recreating economic, legal, and technical institutions, practices, and systems. Industrial technology requires a focus on technology, not just traditional control technologies, but the far more fundamental challenge of integrating environmental considerations into *all* technology and economic behavior.[34]

Sadly, almost all of the evidence paints a compellingly bleak picture from an American perspective.

The Declining U.S. Share of Global Markets

Although the quality of U.S. cars is generally agreed to have improved significantly during the 1980s, the nation's share of the global motor vehicle market remained in decline. In 1984, the North American (U.S. and Canadian) share of global motor vehicle production stood at 37.7 percent.[35] In the space of only six years, it had dropped by over a third, to 27.3 percent in 1990. Most of the difference in American production went to Japanese firms, whose share jumped from 24.1 percent to 32.2 percent. West European auto makers also claimed a chunk of the American market losses, however, as their share climbed from 27.0 percent in 1984 to 30.9 percent in 1990.[36]

Although some analysts attribute these declines to the superiority of, say, Japanese automobiles, many of their features were developed or adopted in response to environmental mandates.[37] Having four valves per cylinder simultaneously improves both fuel economy and performance for only nominal cost, for example. Found on virtually all 1990 models of Toyotas and a large proportion of other Japanese cars, the four-valve systems were rarely employed on "Big Three" cars. Other technologies that were adopted in large measure for environmental reasons but enhanced the performance, durability, or efficiency of cars include on-board computers, radial tires, and lockup transmissions.

Similarly, the competitiveness of U.S. suppliers of power-generating equipment and services also declined throughout the

1980s. According to RCG/Hagler, Bailly Inc., a Washington, D.C.–based firm that provides consulting services to clients ranging from the World Bank to the U.S. Department of Energy, the U.S. share of world exports in this field plummeted by about 50 percent in the period from the late 1970s to the late 1980s.

U.S. exports of power boilers decreased from a high of $1.5 billion in 1980 to less than $250 million in 1988 and 1989 combined. Similarly, export sales of steam and gas turbine generator sets reached a low between 1986 and 1988. Between 1986 and 1989, U.S. exports of power-generating equipment to developing countries ranged between $2 billion and $3 billion; Japanese exports to these countries exceeded $3.5 billion during the same period. In 1989, U.S. exports represented only 10 percent of total exports to developing countries; they had accounted for over 17 percent in 1982–83 and over 20 percent in the late 1970s.[38] Again (as details in Chapter 7 will show), some of the critically desirable attributes of modern power systems were developed in response to environmental requirements.

The United States has lost preeminence in the production of solar cells for producing electricity from sunlight, as well as in fuel cells for generating electricity chemically rather than mechanically, as will be discussed in later chapters. Dominance in wind turbines is shifting to Europe. As recently as 1986, the wind turbines in three of California's mountain passes accounted for 95 percent of the world's wind-generated electricity. But with all of the major powers in Europe having made commitments to take action to curb global warming, installations of wind power there are, in the words of one expert, "unleashing a storm of wind turbine installations" that will likely eclipse U.S. efforts by 1996.[39]

Many of these losses of U.S. preeminence—and the gains elsewhere—are unquestionably related to inferior U.S. environmental standards.

In some cases, the consequences can be described only anecdotally, as in Joel Makower's *The E-factor: The Bottom Line Approach to Environmentally Responsible Business*:

> In the mid-1980s, legislators in both Germany and France enacted zero-discharge laws, effectively banning further phenol dumping

practices. That left the company with the expensive choice of either hauling the stuff to another company or incinerating it. "The plant manager in Germany said, 'Let's look at what's in this waste stream,'" says a former company executive. "And he managed to turn that waste stream into a low-grade phenolic resin, which was sold to industries that were not very sensitive to tight specifications. For them, this material was very acceptable. So, he took a waste stream and turned it into a profit. The plant in France did the same thing. And when the South American plant recognized that the European plants were making money, they started doing this, too."

Ironically, the U.S. plant hasn't managed to change the way in which the resins are disposed, and is paying hundreds of thousands of dollars a year for the privilege of dumping the resins into wastewater. The U.S. operation is stymied because the division that creates the phenolic waste is not the same one that would be responsible for selling the phenol byproduct. Without leadership from the top management, it's unlikely that the two visions will ever come to a meeting of the minds.[40]

Wind turbines, solar cells, fuel cells, and the new coal-burning technology known as Cool Water, or IGCC, are all illustrations of the consequences of lax U.S. environmental requirements, as is the lapse of U.S. leadership in add-on pollution controls.

Indeed, America's once dominant position in air pollution control equipment has deteriorated so much that one investment firm, the Global Environment Fund, has predicted that if U.S. utilities chose to comply with the 1990 Clean Air Act amendments by using scrubbers "it is likely that European and Japanese companies will be the big winners."[41]

Germany is now the world's leader in pollution abatement equipment generally and is particularly strong in the water treatment sector, according to the Organization for Economic Cooperation and Development (OECD). Japan is the major exporter of air pollution control products. While the United States retains its preeminence in waste management, the one environmental field where it has had, until recently, the world's toughest laws,[42] Germany's take-back laws and Japan's recycling requirements will likely result in erosion of the American advantage.

Another consequence of declining U.S. competitiveness is what one commentator calls technology friction, which is "rapidly emerging as a major issue between Japan and the U.S., nudging aside trade friction as the major sticking point between them." This is because technology is the lifeblood of the modern corporation. It begins with research, which leads to the development and production of commercial goods that in turn generate the profits required to recover research costs and sustain further study. If the costs of research are borne by U.S. firms, but the profits are reaped by their Japanese competitors, there can be only one conceivable outcome: eventually American innovation, productivity, and income will decline, while those of Japan will rise.[43]

These trends pose a threat not merely to a few specific industries but to the entire U.S. economy, for there is an emerging body of evidence clearly establishing a correlation between environmental protection and economic growth.

Many U.S. businesses complain that rigorous environmental management impedes economic growth and development. Yet when Stephen Meyer, a professor at the Massachusetts Institute of Technology, analyzed a wide range of state and federal data from the 1970s and 1980s, the results "clearly and unambiguously" refuted such claims.[44]

"In fact," Meyer contended, "the converse was true: states with stronger environmental policies consistently out-performed the weaker environmental states on all the economic measures."[45] In a later study he wrote,

> I expected to find a small to modest *negative* association between environmentalism and economic prosperity [that] would have confirmed the prevailing wisdom at the time [but] it simply was not true. What was truly puzzling was that the analysis uncovered a consistent and systematic *positive* correlation between stronger state environmentalism and stronger economic performance across four of five indicators. States with stronger environmental standards tended to higher growth in their gross state products, total employment, construction employment, and labor productivity than states ranked lower environmentally.[46]

Although Meyer was unwilling to conclude that the positive association between environmentalism and economic growth existed because stronger environmental regulation stimulated economic prosperity,[47] other studies have reached similar conclusions. As one commentator observed,

> Significantly, recent major empirical studies unanimously reject the hypothesis that there is a negative relationship between environmental protection and economic growth. In fact, when statistically significant relationships are found, they are invariably positive; in other words, the U.S. states and nations of the world with more stringent environmental regulation show the best economic performance.[48]

Sometime during the 1980s, the Germans, Japanese, and most of the United States' other industrial competitors reached this conclusion and began reshaping their industries accordingly. Yet the United States continues at an unfaltering pace down the same path that it began following as the 1970s merged with the 1980s. The question that must be asked—and, more importantly, answered—is why.

PART II

LOSSES AND POSSIBILITIES

Chapter Four

U.S. Policy Failures

A business corporation is organized and carried on primarily for the profit of the stockholders. The powers of the directors are to be employed for that end. The discretion of the directors is to be exercised in the choice of means to attain that end, and does not extend to a change in the end itself, to the reduction of profits, or to the nondistribution of profits among stockholders in order to devote them to other purposes.

—DODGE V. FORD MOTOR COMPANY, 1919

The evidence that Germany and Japan are moving much faster to position their industries to capture the benefits of environmental technology leads to an important and troubling question: What do they know that we don't?

The answers are important. Some people are unlikely to accept the reality of the differences among nations without knowing the reasons behind them. Indeed, it is counterintuitive to find highly competitive nations and business leaders acting so differently. The explanations for these outcomes will therefore provide a better understanding of the true dimensions of the problems confronting the United States and, most importantly, of the obstacles that must be surmounted if we are to define and implement solutions.

There are several factors at work. First and foremost are the effects of a large domestic energy industry in the United States. For all practical purposes, Germany produces only coal and Japan imports virtually all its energy. As a result, their industries are attuned to opportunities to lessen expenses, as well as the uncertainties attendant upon dependence on other nations for fuel, by saving energy or developing alternative energy sources. Politically, these options are more viable in Japan and Germany than in the United States because there is no all-powerful constituency of oil, coal, and (to a lesser degree) natural gas interests arrayed against efforts to alter energy use.

In the United States, however, because of the vastness of the land and the historical abundance of inexpensive energy—first wood, then hydropower and coal, and most recently oil and natural gas—domestic industries are among the world's most energy intensive. Many state and local governments derive large fractions of their tax and other income from energy and energy-related industries. In energy-rich states from West Virginia to Texas, severance or other taxes on oil, natural gas, and coal production support public schools, pay for roads and hospitals, and are even laid aside against those "rainy days" that befall even governments. These governments are no more in favor of jeopardizing their relatively stable sources of income than the energy corporations themselves. The result is a powerful domestic political constituency favoring policies that support continued growth in energy production and use or, at the very least, the status quo. Its members include not only the coal and oil industries, but manufacturers of cars, refrigerators, and a wide variety of other goods, as well as state and local officials, university presidents, and the many other beneficiaries of the income stream spun off by energy production, transportation, and use. Investment markets, intent on assuring short-term profits even at the expense of longer-term prosperity, worsen the situation by refusing to finance projects with longer payback periods, such as solar, wind, conservation, or alternative energy projects.

Thus, rightly or wrongly, businesses and many governments perceive their continued vitality—indeed, their very survival—to be contingent on a stable supply of cheap energy. Change—almost any change—is viewed as a threat. And, in all candor, they may be right when American interests are viewed from the narrow perspective of one specific entity, whether it's an energy corporation, labor union, city, or even state. But the interests of the nation as a whole are distinct from those of a particular region. In the longer and larger view of a nation seeking to compete in a changing global economy, those who are unable to adapt to new demands and evolving circumstances are condemned to an ever-declining competitiveness and standard of living. It is the role of the national political system and its leaders to rise above narrow interests and adapt U.S. institutions to meet new challenges. Clearly, that is what has happened in Germany, Japan, and most other industrialized nations, but not in the United States. The explanation for this failure of national leadership lies in the unique nature of the U.S. political system.

The greatest strength of the U.S. political system—its openness—is also its greatest weakness, because this openness makes the government of the United States more susceptible than that of any of the other great democracies to manipulation, especially when those efforts are directed towards averting change rather than facilitating it. Distrustful of government, the nation's founders crafted a system of checks and balances that makes the adoption of change difficult even in the most favorable circumstances. The twists and turns of the U.S. system and the many hurdles that a proposal must overcome before becoming a law all favor those seeking to prevent change. Using tools that range from sham science, used to destroy the credibility of adversaries, to payments of money in the form of campaign contributions, opponents of change can cynically and easily halt all but the most clever and tenacious adversaries. (To completely inventory these tools and describe the manner in which they are deployed is the subject of another book, but for our purposes the review contained in this chapter will do.) Thus armed, those who oppose change in the status quo are able to successfully thwart efforts aimed at making the U.S. economy more efficient and hence more competitive.

These and other factors explain why U.S. corporations have so single-mindedly and successfully opposed policies that are manifestly in the public interest over the medium and long term. They also explain why the Congress and U.S. presidents—elected to safeguard the interests of America and its people, as distinct from those of its corporations—have failed to do their duty. U.S. business and government are clearly capable of taking a longer view, as the development of weapons systems decades in advance of their actual need illustrates. The issue is not whether the U.S. has the capacity to take a longer view of its interests, but why that capability is not being exercised. To more fully answer this question, it is helpful to examine the nature of America's political system through the lens of its most powerful single industry: those who produce, refine, and sell oil. It is there that the sources of the widespread illness afflicting some of the most powerful of America's political and business institutions are to be found.

The U.S. oil industry is one of the greatest concentrations of wealth in the world, while at the same time it is a richly diverse collection of large and small political interests spread through literally every state and city in America. At one extreme is a concentration of enormously wealthy companies, most of which originated when President Theodore Roosevelt dedicated himself to "trust-busting" and shattered one of history's richest firms, Standard Oil, into a handful of other companies ranging from Exxon to Mobil. At the other extreme are tens of thousands of small businesses which pump gasoline, deliver home heating oil, repair furnaces, and deliver a variety of goods and services to virtually every American on virtually every day. When mobilized, these allies—the fabulously wealthy oil companies on the one hand, shoulder-to-shoulder with oil jobbers, street-corner mechanics, and service-station dealers on the other—represent one of the nation's most formidable political forces. Together, they have won victory after political victory over the past several decades. It is an industry steeped in the tradition of influencing politics as first practiced by John D. Rockefeller, founder of Standard Oil, some of whose dealings were described by one historian as follows:

> "Our friends do feel that we have not received fair treatment from the Republican Party," wrote Rockefeller when forwarding a contribution to the party in Ohio, "but we expect better things in the future." But Standard Oil did not stop with contributions. It put the Republican Senator from Ohio on a legal retainership—his fee in 1900 alone was

> $44,500. And it considerately made loans to a powerful Senator from Texas . . . who needed money to pay for a six-thousand-acre ranch he had purchased outside Dallas.[1]

Some are inclined to view the flexing of such political muscle as an artifact of a more corrupt past. But these same political forces continue to be deployed as the nation approaches the twenty-first century. In the twelve Northeastern states from Virginia through Maine, for example, efforts by governors and their chief environmental officers to adopt requirements for zero-polluting and ultra-low-polluting cars have been fought to a standstill by a coalition that includes not only the auto makers, but the oil companies, oil jobbers, and service station operators (see Chapter 6 and the Epilogue).

As of 1986, twelve of the fifty largest industrial companies in the United States were oil companies. Ranked by sales, oil companies accounted for nine of the top twenty-five, and for 30 percent of net earnings. Exxon, the successor to Standard Oil, had 1986 revenues in excess of $69 billion. This concentration of wealth continues to be closely related to political power.[2] As Professor Franklin Tugwell of the Claremont Graduate School explained in 1988,

> To back up its representational and analytical work, the industry has relied heavily on its financial resources. Indeed, its contributions (legal and illegal) have been so extensive that in some instances they have proved a liability—as when it was revealed, in the aftermath of the Watergate scandal, that oil companies provided, directly and indirectly, fully 10 percent of Nixon's reelection funds. Testimony associated with that scandal revealed, again and again, a degree of penetration of the democratic political process that was startling to many observers. . . . In a significant number of cases [in the 1970s, following the first oil crisis], energy companies broke the law in their efforts to influence the newly politicized environment in which they had to operate, handing out large amounts of cash surreptitiously or making other illegal contributions to decision makers.[3]

The oil industry's political influence can be measured in various ways, beginning with numerous tax provisions and other government subsidies (allowances for "depletion" and "intangible" drilling costs alone provided the industry on the order of $50 billion). Many of the most powerful national political leaders of the century were highly dependent on support from the oil industry, including Lyndon Johnson, Sam Rayburn, and Everett Dirksen.

The energy industry also attains political influence by its large role in many state economies. Coal is mined in all parts of the country except New England. Ten states (Alabama, Colorado, Kansas, Kentucky, Montana, New Mexico, North Dakota, Ohio, Tennessee, and Wyoming) collect severance taxes on coal amounting to roughly $500 million a year. Eight states collect such taxes on oil and gas production, providing $3 billion in 1987. In Wyoming, taxes on the energy industry provide roughly two thirds of the state's total revenue.

The economic interests of utility companies also are closely intertwined with those of the fossil fuel industries. Electric utilities frequently own coal and gas reserves or acquire them under long-term contracts to provide a reliable supply at a predictable price. As a result, these industries often maintain a united front in opposition to environmental laws and regulations. (The gas industry is sometimes an exception, since the burning of natural gas can emit much less pollution than oil or coal; this benefit, not surprisingly, is used as a marketing advantage.)

Sheer wealth alone fails to fully explain the enduring ability of these industries to frustrate the enactment of policies that are manifestly in the public interest. The political system provides many other opportunities for industry to exert influence through the Congress.

Consider two stories that ran in the *Washington Post* on May 21, 1993. The first article, headlined "Sen. Boren Targets Clinton Energy Tax," described the efforts of a pivotal member of the Senate Committee on Finance to undo a tax on energy which had been proposed by the recently elected president. That Senator David Boren, a Democrat from Oklahoma, might oppose the Clinton proposal was not particularly surprising. The oil industry, which was bitterly resisting the Clinton tax, accounted for nearly 35,000 jobs in Oklahoma (only Texas and Louisiana had more).[4] Moreover, since February of 1982, the U.S. oil industry had, according to its own statistics, lost 450,000 jobs, while national employment grew by more than 20 million.[5] An oil industry fact sheet describing the losses and complaining of "short sighted government policies" which it said were to blame observed correctly that "oil is the primary fuel source in the United States. It powers virtually all of the nation's transportation system, heats millions of homes, generates electricity and fuels factories."[6]

Boren was joined in his opposition to an energy tax by the chairman of the Senate Committee on Energy and Natural Resources, Senator Bennett Johnston of Louisiana (where 45,300 people are employed by the oil industry). The Republican cosponsor of Boren's measure, Senator John Danforth, wasn't from an oil state, but from a coal producer, Missouri. Not large in the industry by the standards of giants like West Virginia and Wyoming, Missouri was nevertheless one of the eighteen largest coal-producing states, and over one third of its production was exported.[7] In addition, Missouri depended heavily on the U.S. motor vehicle industry, which accounted for one of every fifteen payroll dollars in the state and which would be burdened by the ensuing rise in gasoline prices.[8]

The second *Post* article of May 21, 1993, headlined "Fanning a Prairie Fire," revealed the mechanism behind the influence. It described how a sophisticated industry-financed campaign had positioned Boren so that like it or not he had little option but to reject Clinton's proposal. Boren, said the *Post*, "was bombarded by calls and mail to his Oklahoma and Washington offices. Everywhere he turned, he heard the projections of job losses in his oil rich state. And he was facing taxpayer rallies planned in major cities Friday and the release soon of a state public opinion poll showing overwhelming opposition to the plan."[9]

On the surface, Boren's response would seem a textbook example of the way a democracy should work. But there is a difference between genuine voter sentiment and the work product of an adroit, skillful, industry-financed campaign. Because Boren hails from an oil state and is a relatively conservative Democrat, he was a prime target for the oil and gas industry in its campaign to defeat the tax. Rather than diminish the credibility of its arguments by voicing them directly, however, the industry employed two industry-funded surrogates to do its work—Citizens for a Sound Economy

(CSE), which calls itself an educational organization, and the Affordable Energy Alliance (AEA), a coalition of industry firms allied to defeat the tax. Both are based in Washington, D.C.[10]

Citizens for a Sound Economy mounted a $100,000 advertising campaign and organized the two rallies through a direct mail blitz to 9,500 Oklahomans. For its part, with the help of a former Boren aide who was a member of a state energy regulatory commission, AEA commissioned a study of the tax's economic impacts in Oklahoma. The study was done by a component of the Business College of the University of Oklahoma, the Center for Economic and Management Research, after AEA pledged to donate $5,000. AEA also hired a prominent pollster for the Oklahoma Democratic party.

The episode was typical of the manner in which political decisions are shaped in Washington, and illustrates what is both the great strength and great weakness of the United States—that it is a fabulously wealthy democracy. The sheer amount of money involved means that decision makers must take into account the interests of coal, oil, and gas producers.

By contrast, political leaders in Germany and Japan can make decisions relatively unhindered by the political power of "Big Oil" or "King Coal." This is not to say that the industries of Japan and Germany are so weak that their views are disregarded in the political process. On the contrary, both nations systematically incorporate business into the decision-making process. In Germany, for example, government ministries are required to consult with both industry associations and trade unions during the legislative drafting phase. In the view of one U.S. business group, "this consultative process means that German industry associations are generally more powerful actors in the policy-making process than their American counterparts."[11] Even the national research effort, administered through a system of thirty-eight applied research institutes, is tailored to the needs of industry. Only research that is a priority for industry will receive government funding.[12] The relationship between Japanese government and business is less formalized but equally intimate.

However, even though the governments of Japan and Germany take the needs of their domestic industries into account, they also maintain a clear vision of what their national interests require; these decisions—what to do for the sake of the common good—can be made with relative objectivity. In the United States, the very openness of the system of government precludes such rational decision making.

Of course, this openness makes it possible for groups such as the Sierra Club and The Wilderness Society, which represent tens of thousands of voters, to exert leverage over presidents, senators, and representatives. But this potential for lobbyist influence also makes the system subject to manipulation and abuse by corporations, allowing them to magnify their legitimate concerns out of proportion to their true merit. The power of lobbies for the energy and related industries explains policy failures which have become so grave that they threaten the nation's very survival as a viable industrial democracy in a changing world.

In previous eras, money bred corruption of an old-fashioned sort—bribes exchanged for construction contracts, for example. This is still the case in Japan, where in 1993 evidence of political corruption led to the first change in ruling parties in decades. But in Japan, as previously in Washington, money was not employed to distort the process of governing itself. Rational decision making about major policy issues continued, even if an occasional bridge or dam were thrown in the direction of well-heeled contractors.

What's different today is that money is tilled into the soil of Washington like chemicals into a farmer's field, disrupting and distorting the very process of government. The covert regime of money began in the mid-1970s. In *The Suicidal Corporation*, Paul Weaver, a former executive of the Ford Motor Company and a veteran writer for *Fortune* magazine, described the influx of pro-business advocates that forever changed Washington decision making.

> Business's retreat from responsibility was completed as the political activism of the 1960s and 1970s prompted a huge buildup of the Washington business lobby. By 1980, the number of trade associations based in the capital stood at 2,300, up 250 percent from a decade earlier. Some 500 companies had Washington offices. The U.S. Chamber of Commerce boasted over 200,000 individual and institutional members and a budget of over $60 million a year. Half the new downtown Washington office space being built at breakneck speed was being leased by associations, attorneys and accountants. . . . [Business lob-

> bying activities accounted for] about 7 percent of the capital area's total labor force, including military personnel. The main effect of all this growth was to fragment the business lobby and to focus companies' attention more and more on momentary individual interests, and less and less on enduring common interests.[13]

By 1994, the budget and membership of the U.S. Chamber of Commerce had grown only modestly, rising respectively to "just under" $70 million and 219,000, according to Hank Kopcial, director of corporate relations for the organization. The number of companies with Washington offices, however, had skyrocketed, and stood at roughly 7,000, he said in a 1994 interview. Another 4,000 U.S. and foreign firms chose not to maintain Washington offices, but were represented by area lobbyists or lawyers.

The government has become, in short, just another aspect of the market to be manipulated at will, and money—especially in the form of campaign contributions—is the tool through which it is controlled. In 1988, the total cost of all campaigns for federal office (for seats in the U.S. Senate and House of Representatives and for the presidency, including both primary and general elections) was roughly half a billion dollars. That sounds high. But it's less than the cost of pollution controls on just three power plants—and there are hundreds in the United States, not to mention refineries, chemical plants, pulp and paper mills, cement kilns, and thousands of other sources of the billions of pounds of pollution that spew into the air and water and onto the land each year. From a polluter's perspective, it's far cheaper to invest in politicians than in environmental controls.

In the words of Paul Weaver,

> Business views the political system as a source of business advantage. Almost anything can be a business advantage—a subsidy, a tax break, an entry barrier, a spurt in the rate of economic growth, a government purchase, a regulatory move that hurts a competitor, etc. Thus business's agenda has an open-ended, mercurial opportunistic character.
>
> Some [business advantages] are overt . . . but most business advantages, however, are covert. They flow from policies whose overt purpose is to accomplish an unrelated public objective, such as reducing

air pollution. But policies have unacknowledged or unintended practical effects that help some businesses and hurt others. Most business lobbying is intended to capture the covert benefits of public policies and deflect any covert costs onto others.[14]

Still, corporations are managed by men and women just like us. Surely, we ask ourselves, these people are able to recognize a greater good that sometimes requires individual inconvenience or sacrifice. The answer is that the people who manage a corporation's affairs are indeed people like us, but the entity to which they owe their fealty—the corporation—is not.

Corporations are artificial intelligences, created by us as the most economically efficient mechanism for converting resources and labor into goods. Business corporations, unlike the humans who work for them, never grow thirsty or hungry, weary or sleepy. They merely exist, and for only one reason—to make money.

Most people will readily agree with the last statement but without realizing its grave implications, one of which is the subject of this book. This simple truth—that corporations exist solely to make money—will be confirmed by virtually any lawyer or corporate officer. Corporations can be created because of laws that allow them, and those laws uniformly require corporations to concern themselves only with making a profit. Thus, to the corporation itself money is more important than the quality of the environment or the integrity of the American political system.

The sheer force of this fact can be illustrated by an encounter between Henry Ford and corporate law about seventy years ago. Although Ford became known late in his life as a ruthless adversary of organized labor and a conservative of such extreme views that he was rumored to have flirted with Nazism, his early career was in some respects a model of selflessness and compassion: he introduced the eight-hour day and the five-dollar daily wage, instituted a profit-sharing plan for his workers, and supported efforts by pacifists to halt World War I through mediation.

Ford's base of power and wealth was the Ford Motor Company, which he founded in 1903. It began production in 1909 of what is probably history's most famous car, the Model T, popularly known as the "flivver." A standardized vehicle produced on an assembly line, the Model T's low price brought automobile ownership within the reach of millions of middle-class Americans.

By 1919, Ford sales had been so extraordinary that the company was declaring not only 5-percent monthly dividends, but year-end and special dividends as well. At a price of $440, the car was selling so well that Ford, then owner of 58 percent of the corporate stock, decided to expand production. "My ambition is to employ still more men," Ford later told a court, "to spread the benefits of this industrial system to the greatest possible number, to help them build up their lives and their homes."

And Ford wanted to spread the benefits of his exceptional success even further. "Ford was of the opinion that his company had made too much money," writes legal scholar Norman Lattin, "and that although large profits might still be made, they should be shared with the public by reducing the price of the car." But when the automotive genius decided to drop the price of a car from $440 to $360 he was promptly sued by the company's minority shareholders and just as promptly rebuked by the court. The court barred the price cut and admonished:

> There should be no confusion (of which there is evidence) of the duties which Mr. Ford conceives that he and the stockholders owe to the general public and the duties which in law he and his co-directors owe to protesting, minority shareholders. A business corporation is organized and carried on primarily for the profit of the stockholders. The powers of the directors are to be employed for that end. The discretion of the directors is to be exercised in the choice of means to attain that end, and does not extend to a change in the end itself, to the reduction of profits, or to the nondistribution of profits among stockholders in order to devote them to other purposes.[15]

Corporations have evolved considerably since the court's rebuff of Henry Ford's attempt to give the public "a break." Many states have now enacted laws explicitly allowing corporations to act for the benefit of the public, even if that might mean some reduction in profits. But not all states have such laws, and even in those that do, the amount that a corporation can donate to charity is limited. In some cases, the limit may be an explicit amount (5 percent of net income, for example) but it can never be a sum so large as to be "unreasonable." Whether or not corporations owe some duty to "acknowledge and discharge social as well as private responsibilities," this obligation is always subject to "reasonable limits."[16] Profit is the yardstick against which the corporation and its officers are measured.

It is this legal compulsion that explains why corporations are driven to hire literally tens of thousands of lawyers and lobbyists and corrupt the political system through campaign contributions so massive that it's no longer necessary to buy a single vote because the entire institution has already been auctioned off.

That money changes hands to buy a specific vote is not the issue here, because it usually doesn't. Even when votes are bought and sold, however, proving it is well-nigh impossible. Increasingly, decisions are made in secret. The 1990 amendments to the Clean Air Act, for example, were written almost entirely in meetings from which the public and press were barred, as were the final provisions of the 1986 Superfund amendments. Closed doors have become the rule rather than the exception, and requirements to the contrary are simply ignored. With no minutes and no votes there's no evidence, and therefore no proof of political wrongdoing.

The pervasiveness of polluters' money explains a string of broken and unfulfilled environmental promises from Republicans and Democrats, the Congress and the White House alike. In 1989, Senate Democratic Majority Leader George Mitchell of Maine called levels of air pollution in the United States—which experts estimate kill 50,000 or more Americans each year—"a public health emergency . . . requiring prompt and vigorous national action." There was good reason for alarm: smog levels were the worst in a decade,[17] average global temperatures were the highest ever recorded,[18] and even the chemical industry was conceding that the cause of an ozone "hole" as large as North America and as tall as Mt. Everest was pollution, namely the family of chemicals known as chlorofluorocarbons, or CFCs.[19]

Had the 1990 amendments followed the pattern of German law, U.S. emissions of sulfur dioxide, the chief cause of acid rain, would have been reduced by over 14 million tons per year. Comparably large reductions also would have resulted from adopting the Japanese approach of mandating specific technologies. In either case, utilities would have been compelled to introduce new technologies ranging from cleaner methods for burning coal to the Knauf "homes-from-pollution" systems. There was ample evidence of widespread environmental and health damage to warrant sulfur dioxide emission reductions of this level. Instead, Congress chose to mandate a reduction of roughly 9.5 million tons, which can be achieved largely through the simple expedient of burning lower-sulfur coal, combined with the use of scrubbers at a limited number of plants.[20] Not surprisingly, the boom market in air pollution control equipment that many expected to follow enactment of the amendments failed to materialize, leading an industry witness to complain in February 1993, "I'd like to report to you that 28 months after passage of the amendments, the U.S. air pollution control market was smaller in 1992 than it had been in 1991, and that the 1991 level has remained essentially flat for the past ten years at less than $1 billion per year. In short, there has been no real market impact from the amendments."[21]

Of course, it's not the job of Congress to create unnecessary environmental requirements merely for the sake of sustaining the pollution control industry. Just as clearly, however, it's not the job of Congress to *avoid* imposing those requirements when they are manifestly necessary. Yet that's precisely what happened in the case of the 1990 amendments and, increasingly, in the case of other energy and environmental legislation as well.[22]

Under Mitchell's guidance, Congress proceeded to pass a 328-page set of amendments to the Clean Air Act which established a new federal right—not for Americans to breathe clean air, but for industries to pollute. Another amendment expressly legalized the continued production and use of chemicals that destroy stratospheric ozone until the year 2030, even though 2 to 4 percent of the global ozone layer had already been destroyed. Yet another provision flatly prohibited U.S. regulation of lead, a toxic metal that destroys intelligence in children and is associated with 50,000 deaths per year from heart attacks and strokes, as a hazardous air pollutant. The meetings at which these provisions were crafted were held behind closed doors. No minutes or other written records were kept. As a result, the 1990 amendments continued the United States on the path of increasingly lax environmental requirements calculated to avoid the need to develop new technologies.

The deft use of bogus citizen groups, as seen in the Boren case, is common. In the case of acid rain, the coal and utility industries relied on a group called Citizens for Sensible Control of Acid Rain, a coalition of about 125 coal and electric companies. It doled out more than $5 million during the 1980s, and in 1986 spent more on lobbying than any other organization in Washington. Though its name suggests a broad-based grass-roots organization, not one of the "citizens" that were members were real humans—all were corporations. The same can be said for the hundreds of other sham organizations that purport to represent citizens or constitute "alliances." The following are a few examples.

The Marine Preservation Association, a nonprofit organization created and funded by about fifteen major oil companies ranging from Exxon to Texaco, funnels money into efforts to repeal oil-spill liability laws in states from Maine to California. MPA sounds as if it's dedicated to protecting the environment, but its articles of incorporation say MPA is "organized exclusively to promote the welfare and interests of the petroleum and energy industries."

The National Wetlands Coalition, a group that sports the eco-sensitive logo of a waterbird in flight over cattails and marsh grasses, is "at the forefront of an aggressive effort to rein in federal wetlands rules," according to the *Washington Post*. With an annual budget of $400,000 and a membership list that includes Exxon, Shell, Texaco, and a wide range of real estate interests, the coalition is seeking multibillion dollar buyouts from federal taxpayers as the price of allowing swamplands and marshes to remain undeveloped.[23]

Responsible Industry for a Safe Environment (RISE), a half-million-dollar-per-year lobbying group supported by pesticide manufacturers is dedicated to killing proposals that would require lawn care companies to post warnings after spraying chemicals that can cause cancer and nervous system damage.[24]

The Safe Buildings Alliance, a group of three former manufacturers of asbestos, peddles the feel-good message that "you have more of a chance of being hit by lightning than dying from asbestos," in the words of its vice president, Jeff Taylor. Never mind that some experts put the numbers of Americans killed by exposure to the cancer-causing mineral at 100,000 over the last several decades.[25]

Virtually unheard of before 1980, such groups as these are now routinely created on a regular basis by the high-powered lobbyists and public relations firms hired by polluters to fend off new laws, cloud scientific findings, and quell public outcry.

An official of the Federal Election Commission estimates that there are over a thousand such lobbying groups, each organized around a single issue or ideology—and it's often impossible to tell what a group believes in from its name. The head of one Washington-based public relations firm told the *New York Times* that he had previously organized "a number" of such "phony coalitions," but abandoned the practice. It was too "personally disturbing," he said.[26]

One of the most amusing—and alarming—of these groups was first revealed in a full-page newspaper advertisement displaying a donkey sporting ear muffs and a scarf, highlighting a large-typeface question: "If the Earth is getting warmer, why is Minnesota getting colder?" It was a seemingly good question asked by an apparently reputable group, the Information Council on the Environment, or ICE for short. The trouble is, both the premise behind the question and the group asking it were, once again, little more than fronts for corporate polluters. In this case, the "council" was a public relations and advertising agency operating with coal and electric company money to "reposition global warming as theory (not fact)," in the words of an internal document.[27]

The hook question was based on the spurious claim that Minnesota has cooled. There *was* a dip in temperatures from about 1820 to the 1860s, but since then, except for a slight dip in the 1950s, the state has warmed steadily. It's now about 3 degrees Fahrenheit hotter than in 1860, according to researchers at the University of Minnesota. Such facts notwithstanding, the coal and electric industries flooded the airwaves and saturated the newspapers of three cities (Bowling Green, Kentucky; Flagstaff, Arizona; and Fargo, North

Dakota) with variations of the ad to test the effectiveness of their message. Before the ICE campaign had a chance to take hold, however, the group was exposed by articles in both the local papers and the *New York Times* and it quickly vanished from view.[28]

Such campaigns cost money (the tab for the three one-week test marketings of the ICE ad was put at $500,000 or more), but the evidence suggests they're worth the price from the polluters' perspective. In just the last three years, for example, there has been a decline in the apparent level of scientific concern over threats ranging from global warming to dioxin. In virtually every case, a high-powered public relations firm can be found lurking in the background, quietly undermining the credibility of reputable scientists and creating the appearance of scientific disagreement when the dissent within the mainstream of researchers is slim to none. In the case of ICE, for example, names of three scientists were given to reporters to validate claims of cooling. But when contacted by reporters two of the three renounced their connections to the ICE campaign.[29] Similarly, critics of government programs to eliminate lead in drinking water, paints, gasoline, and other products cite recent claims by two scientists that basic studies of lead's health effects are "seriously flawed." What they fail to mention is that the two scientists making that claim have been paid to testify at judicial proceedings on behalf of the lead industry.[30]

In contrast, the major environmental organizations depend largely on their own members for support. The National Wildlife Federation, for example, has nearly a million members, while the Sierra Club raises the lion's share of its $35-million annual budget from $33 per year in dues contributed by each of its 560,000 members. They and other environmental groups sell calendars, posters, and even sweatshirts to make ends meet, while the so-called alliances and coalitions derive their cash almost exclusively from corporate coffers.

Often the hands on the keys to the corporate cashboxes are those of people who themselves were once the objects of such manipulation—not senators and representatives whose names appear in the headlines, but little-known and extraordinarily powerful congressional aides—"staff" in the parlance of Washington. "Unelected representatives," they were called by author Michael Malbin. It is these men and women, first while serving members of Congress, then later as lobbyists, who often determine the fate of environmental legislation. Hired by the thousands, these former insiders fill the hearing chambers, courtrooms, halls, offices, and restaurants of Washington, peddling information and influence.

Bill Fay, who headed the industry lobbying alliance aimed at influencing amendments to the Clean Air Act, was one such product of the revolving door. As head of the 2,000-member Clean Air Working Group, Fay modestly described his contributions as merely bringing "an understanding of the process" to the debate over the clean air amendments. But one of Fay's former colleagues took a different view: "What he did was raise money and scare the hell out of the U.S. Senate. There's no doubt that he coordinated hand-in-glove with Senator Symms [Republican Steve Symms of Idaho, Fay's former boss and the principal Senate opponent of amendments to strengthen the Clean Air Act] and the dozen other senators who were trying to throw a monkey wrench into the process."[31]

Bill Fay enjoyed experienced allies in his struggles over clean air. His two right-hand men were a former counsel to the Republican members of the House Committee on Energy and Commerce and a former aide to John Dingell (D-Michigan), chairman of the committee. These three produced an industry effort with "a level of sophistication never seen before," in Fay's words. "Were there meetings?" he asked in response to a question about the industry relations with members of Congress. "A lot of them," he replied. "[The] oil industry [was] meeting with oil states, the steel industry with steel states," and so on. "I encouraged that," he added. "We worked with anybody that would let us work with them."[32]

It is not just Congress that contributes to the problem of the revolving door. For example, Hank Habicht was the U.S. Justice Department's chief prosecutor of Superfund crimes during the Reagan administration. He left in 1987 to join an industry-backed coalition including Aetna, Union Carbide, Dow Chemical, and DuPont to develop amendments to Superfund legislation. His tenure with the industry was cut short by his return to government, where he then served as the second-ranking official at the Environmental Protection Agency. One of his responsibilities: Superfund.

Habicht's partners in the Coalition on Superfund, as the industry venture was called, included William D. Ruckelshaus. Ruckelshaus was the first administrator of the Environmental Protection Agency (EPA), after which he served as head of environmental affairs for the giant forest products corporation Weyerhauser. Then it was back to EPA. From there, Ruckelshaus founded the Coalition on Superfund. Then he took over the presidency of Browning Ferris, Inc., the nation's second-largest handler of hazardous waste. When Ruckelshaus and Habicht left the Superfund coalition, control of the organization remained in the hands of another user of the revolving door, Lee Fuller. A former staff director of the Senate Committee on Environment and Public Works and onetime chief environmental aide to Senator Lloyd Bentsen of Texas (later President Clinton's Secretary of the Treasury), Fuller's other clients included the nation's third-largest waste handler, JZC, Inc.

Perhaps because of the national outcry over toxic waste, such companies seem to have a special affinity for once-powerful aides. For example, Waste Management, Inc., the nation's largest waste company, directly or indirectly employs a former presidential aide, a former assistant to the Senate majority leader, a former EPA general counsel, two former lawyers for major environmental groups, a former aide to Senator Edmund Muskie, and a former acting administrator of EPA, to name but a few.

Such men and women are professionals who sell their skills to the highest bidder. Now that the fight over clean air is over, for example, Fay has moved on to head a coalition aimed at "reforming"—the euphemism that lobbyists usually use for gutting—the nation's product liability laws. Another lobbyist, Kevin Fay (no relation to Bill Fay) headed the Alliance for a Responsible Chlorofluorocarbon Policy, the industry group that successfully staved off U.S. regulation of the ozone-destroying chemicals for roughly a decade. Now that CFCs are being phased out, he has gone on to head the Alliance for Safe Atmospheric Policy (ASAP), an industry group representing companies with a stake in the debate over global warming.[33]

The rough numbers (exact ones aren't kept) suggest that the flow of influential staff leaving the Congress to work as corporate lobbyists is approaching a flood. Of the roughly 16,000 congressional staff members, Wayne Walker, editorial director of Staff Directories, Ltd., classifies 3,200 as "key." An indication of the escalating turnover among these aides is that in 1989, after thirty years of being published only once a year, the *Congressional Staff Directory* was converted to a twice-a-year schedule. "You'd look at the copy and say, 'My God, these names have all changed,'" he added. Most of the departing staff, he believes, end up working as Washington lobbyists.[34]

This view is confirmed by a check of the thirteen key environmental staff who left the Senate Committee on Environment and Public Works between 1984 and 1990. Of the thirteen, twelve went to work as lobbyists for industry. They or their firms represent not only virtually every industry with a stake in environmental laws but dozens of individual corporations, including Union Carbide, GPU, and Exxon—the companies responsible for Bhopal, Three Mile Island, and the Exxon Valdez. The number who joined public interest groups? Zero.

Hiring former congressional staff assures that an industry's representation will be "more effective, more efficient, and more likely," in the words of one congressional aide. "They get their position represented and a voice in shaping the final product." But the expertise of these men and women, when brought to bear on highly technical subjects for the purpose of protecting special interests, produces other results as well: sometimes bad laws and invariably delays. "The Congress has come to a standstill—it can no longer legislate," is the conclusion of Wayne Walker of the *Congressional Quarterly*, who places much of the blame on the revolving door.[35]

Presidential Initiatives

Powerful as it is, Congress is only one branch of the U.S. government. The executive branch deserves much of the blame for America's technological decline since 1980. As Japan committed itself to a third round of cutting energy consumption and Germany awakened to the ravages of air pollution, America elected Ronald Reagan to the presidency. To his credit, Reagan lived up to his promises. Having pledged to cut the budget, lessen federal regulation, and commit the nation to a free market economy, he did so. The consequences, although devastating, did not begin to manifest themselves unmistakably until the decade's end.

The need to cut back on environmental regulation had been one of President Reagan's campaign themes. He had called for returning the primary responsibility for environmental regulation to the states. "They've got rules that would practically shut down the economy if they were put into effect," David Stockman remarked of the EPA soon after the election. As director-designate of the Office of Management and Budget, Stockman wrote an apocalyptic memorandum warning that the recent buildup of environmental and safety regulations would "sweep through the industrial economy with near gale force, preempting multi-billions in investment capital, driving up operating costs and siphoning off management and technical personnel in an incredible morass of new controls and compliance procedures."[36]

Among the principal casualties of these abrupt policy reversals were the products of American environmental and energy ingenuity and the rules that encouraged them. Tighter miles-per-gallon standards for automobiles were killed, as was a program for increasing the efficiency of refrigerators, air conditioners, and other large appliances. Though the military budget doubled, federal spending for conservation, solar energy, and other non-oil programs dropped by two thirds. Price controls on oil and gas were lifted, while the federally funded search for oil substitutes like oil shale was killed. Even the solar collectors that provided the White House with hot water were removed.

For the next twelve years, money and staff for energy and environmental regulatory programs were slashed.[37] The Environmental Protection Agency was systematically stripped of its independence, its authorities transferred by a series of executive orders to two arms of the White House.[38] As vice president, Bush presided over the President's Task Force on Regulatory Relief, to be succeeded by his own vice president, Dan Quayle, who chaired a successor group, the Council on Competitiveness.

Funding for research and development, too, was slashed, again for the ostensible purpose of maintaining a free market. One business consultant described the attitude that he confronted at a meeting with a senior official of the Office of Management and Budget:

> I was amazed at the lecture I received, unsolicited, from an OMB official who quoted Adam Smith, literally, to tell me that the only legitimate research functions of government were for defense and navigation. He allowed as how the government would probably continue to fund some basic civilian research, though many in OMB also felt this was inappropriate and a waste of money. When I asked him about international competition, he was clear and definitive in saying that if foreign governments were, in fact, assisting companies with applied research and development, it was because they were misguided socialists whose economies would suffer because of too much government interference. The marketplace on its own could spend what was needed for R & D.[39]

These cuts devastated efforts in the United States. Consider, for example, the U.S. government's funding for conservation. In fiscal year 1983, President Reagan's budget request for conservation research and development—programs that had yielded superefficient light bulbs and refrigerators—was zero. Congress continued funding these conservation programs but at levels far below those of the fiscal years 1979 to 1981. Indeed, in the space of one year, 1982, funding was slashed by 71 percent.[40] Whatever the merits of various arguments that too many tax dollars are wasted by federal bureaucrats in one program or another, the plain truth is that some of these dollars are not wasted and cuts of that magnitude simply cannot discriminate between the good and the bad. Every program suffered, and the story was the same almost everywhere. Only the coal programs managed to survive somewhat unscathed, thanks largely to their powerful allies in Congress.

As a direct result of the actions of the Reagan and Bush administrations, America fell from dominance in the solar industry (recall Stan Ovshinsky's plight at ECD) and began retreating from its efforts to build higher-mileage autos, thus ceding leadership on a wide range of automotive technologies to Japan. The incentives that existed to improve cleaner methods for burning coal and other cleaner and better technologies disappeared, and with them the U.S. lead over Japan, Germany, and the Netherlands (recall the loss of the "homes-from-pollution" process). Canada was able to pick up one fuel cell technology and Japan a second.

The Cheap Oil Doctrine

The list of such U.S. tragedies is lengthy, and many are explored later in this book; however, Reagan's most devastating policy was one largely hidden from public view.

Unannounced, Reagan launched the nation in an utterly new direction in energy policy. For twelve years a strong, consistent, and directed—but nearly invisible—national energy policy returned the United States to a "free market" reliance on cheap Persian Gulf oil.[41] The three presidents before Reagan—Nixon, Ford, and Carter—had all advocated the development of domestic alternatives to Persian Gulf crude oil, a policy rejected by Reagan.

The principal architect of Reagan's energy and environmental policies was the first director of the Office of Management and Budget, David Stockman. He crafted an energy strategy that called for the United States to intentionally *increase* its reliance on Middle Eastern oil and, when necessary, wage war to keep cheap oil flowing. When U.S. Marines were dispatched to swelter in the blistering heat of Saudi Arabia during 1990, it was because of a conscious decision made a decade earlier.

Stockman explained his reasoning in the fall 1978 issue of *The Public Interest*, a conservative policy journal. Stockman pilloried the energy self-sufficiency proposals of the Nixon, Ford, and Carter administrations. They were, he said, "cramped, inward looking" strategies based on "Chicken Little logic." Calling for a strategy of "reliance on the world market for energy," Stockman dismissed fears of OPEC extortion as "economic mythology," saying "The actual geopolitical dangers . . . require only two policies—strategic reserves and strategic forces."

To this day, this cheap oil doctrine remains the real energy policy of the United States. President Clinton proposed a modest increase in energy prices through a new tax but was almost immediately turned back by Congress—so quickly, in fact, that some have questioned whether Clinton himself ever cared or understood what was at stake.

Those supporting the cheap oil strategy—the oil industry, for example—defend it as "allowing the free market to work." But world trade in oil is not a classic free market, in theory or fact. About 75 percent of the world's oil is controlled by the thirteen-nation cartel formed in September 1960, the Organization of Petroleum Exporting Countries, or OPEC. Oil flow is metered by a handful of leaders who are at best uneasy allies of the United States, and at worst outright enemies. The hand on the pump is not Adam Smith's, but the hand of the likes of Ayatollah Khomeni, Saddam Hussein, and Moamar Quadaffi.[42]

In fact, beginning in the early 1980s, Reagan and Bush appointees themselves, acting through Saudi Arabia, helped control both the flow of oil and its price. United States officials repeatedly discussed the price of oil with the Saudis, the largest oil producers among the OPEC nations, emphasizing the negative consequences of prices which were either too high or too low.[43] Thus America's future was entrusted by Reagan and Bush not to a free market, but into the hands of the Saudis.

In the process of increasing its reliance for oil on the global market, the United States was surrendering something vastly more important than money—independence. Make no mistake about it, the Reagan/Bush policy of the 1980s and 1990s—burn imported oil and fight to get it—effectively dictated many other policies.

Oil imports from the Middle East were about 6 percent of total supply in 1973, then peaked at about 18 percent in 1979, dropped, then rebounded to 12 percent in 1989. Thus, in 1989 the United States was twice as dependent on Persian Gulf oil (relatively and absolutely) as it had been in 1973. The 600 million barrels of oil stored in the Strategic Petroleum Reserve was only enough to cushion short-term disruptions.

As long as the United States depended on a fragile lifeline of oil flowing in one direction and dollars in the other, massive military might had to be maintained. Armed force was required both to maintain the steady flow of oil and to counter the military capability of the countries that sold oil. As long as imported oil remained the lifeblood of U.S. cars and trucks, American men and women were required to defend it. Part of the "peace dividend" was being deposited in Saudi Arabian bank accounts.

Investing in fuels rather than fuel efficiency was implicitly a decision to direct capital and profits toward the coal, oil, and gas companies rather than toward the industries that manufacture capital goods ranging from high-efficiency refrigerators and air conditioners to high-mileage cars. But even worse, much of the money went overseas. The annual bill for U.S. oil imports in constant 1989 dollars was roughly $50 billion. Cumulative U.S. payments for oil imports between 1970 and 1989 totaled $1.1 trillion—roughly three years of U.S. defense spending.[44] In short, the policy constituted a huge drain on American capital, impeding investments toward a more competitive industrial infrastructure.

A decision to rely on oil and the automobile was a commitment to continued dependence on streets, highways, and bridges, rather than on rails, rehabilitated housing, and neighborhoods. Urban growth and government spending was caught in a vicious cycle. Worn-out highways had to be repaired because so many people commuted from the suburbs that had been built around the roads; with no money available to rehabilitate urban housing or construct urban transit, workers had to continue to commute by car.

By 1990, only Australia burned a larger share of its energy in cars, trucks, and planes than the United States. The Australian percentage of energy allotted to transportation was 37.9 percent compared to 36 percent in the United States. Even Canada, with more vast spaces and a sparser population, devoted only 25.6 percent of its power to transport, according to data compiled by the International Energy Agency.

Despite the increased fuel economy of cars due to CAFE (Corporate Average Fuel Economy) and other standards, the share of U.S. energy going to cars and trucks continued to climb through the 1970s, 1980s, and into the 1990s as well. After the first oil shock of 1973, oil consumption in the United States declined in virtually every major economic sector except transportation. For example, oil used for the generation of electricity and the heating of buildings was off almost 50 percent. Industrial use was off 10 percent. But transportation demands had jumped by 20 percent, according to the World Resources Institute.[45]

The decision to continue burning large quantities of coal, oil and natural gas was a decision to continue producing the pollutants which cause smog, acid rain, and global warming—roughly 30,000 pounds a year for each man, woman, and child in the nation. About one out of every five pounds of carbon dioxide emitted in the world originates in the United States—more per person or per unit of gross national product than in any other industrialized nation on earth.[46]

Producing oil scars the land. Shipping it pollutes rivers, harbors, and oceans. Burning it degrades the air. Mining coal leaves open or thinly bandaged sores, oozing pollution into wells and rivers. Addiction to both condemns America to a future of a "catch-up" policy, constantly chasing environmental harms rather than preventing them at the outset.

Yet even the thinnest pancake has two sides, and even as Ronald Reagan and George Bush were turning the clock back in Washington, their counterparts in California were looking toward the future. There, policies were being developed that would help cushion the blows dealt to national competitiveness by Washington.

Chapter Five

California Sunshine

> The other forty-nine states may be willing to lose their technological edge to the rest of the industrialized world, but not those of us who call California home. Air pollution is not only a threat but an opportunity.
>
> —LARRY BERG,
> South Coast Air Quality Management District

In the predawn darkness of Newport Beach, a city bus roared down Newport Center Boulevard, past the manicured lawns and sprawling upscale shopping malls of the community that some Californians call America's Riviera. Californians are different from other Americans, and it showed that morning. California was in the grips of a water shortage, but water seeped from underground sprinklers and trickled from dense green turf. It darkened the concrete sidewalks and left the asphalt glistening. The bus's tires hissed as they sped through the slicks. On nearby highways, headlights glared as thousands of cars sped through the lush richness of Orange County, carrying mostly white, mostly rich commuters to their skyscraper offices ninety miles away, near the barrios and slums of Los Angeles.

Everybody in California seems to want to know the time— and to the split-second, not merely the minute. The hosts of *Los Angeles Today*, the city's popular drive-time television news program, announced that it was precisely "six-oh-seven and a half." The Golden

State Freeway was already tied up by an overturned car. A stalled bus blocked the Pomona Freeway. A three-car collision snarled the Pasadena Freeway. An energetically cheerful weatherman declared that today, October 1, 1992, would bring a welcome end to the fierce hundred-degree heat which had blanketed Southern California for two weeks, creating a smog inversion stretching from Hawaii to San Bernardino.

Other lights began to appear. They winked on through the windows of the Newport Beach Marriott as businessmen (and a few businesswomen) prepared for the second day of a meeting to discuss opportunities. But here again, the Californians were different, for the focus of their discussions was not the opportunities created by the state's multibillion-dollar aerospace or silicon chip industries, but the profit and employment potential generated by the near-lethal levels of smog for which the state is known throughout the world.

There was a curious optimism among the 310 men and women. Though the state's unemployment level hovered in the double digits (the worst since the Great Depression), and despite a decline in tax revenues and personal incomes which threatened the serenity of America's paradise, a palpable enthusiasm filled the meeting rooms. Talk dwelt not on the dismal present, but rather on a bright future: how El Pollo Loco's new and nearly pollution-free systems for grilling chicken might be marketed in Japan and Europe; whether Alzeta's unique "pyrocore" burners could solve pollution threats in Germany and the Netherlands, and thus create new jobs for unemployed Californians; how soon the new battery-powered cars required by state law would be on the road and how well they might sell in the other forty-nine states.

Speakers occasionally mentioned—as virtually every Californian does repeatedly to visitors—that California's economy would rank as the world's eighth largest if only it were a nation. Every speaker, and every attendee, was focused on sustaining California's economy and expanding it if possible. They planned to do this by marketing the technologies in which California clearly leads the United States and in many cases the rest of the world as well, those that protect the environment.

Californians think of themselves as separate from the rest of the United States, and often the feeling is mutual. Its policies are so different from those set in Washington, D.C., and the other forty-nine states that it is, at least where energy and the environment are concerned, a different world. To attribute California's successes in the 1980s to the rest of the United States is misleading, for the state's victories have often come despite federal policies, not because of them. When Washington abandoned solar and wind power, for example, California nurtured them; when Washington weakened tailpipe emissions standards, California strengthened them; and when Washington turned its back on conservation, California embraced it. As a result, California makes more electricity from sunlight, wind, and heat from the ground than any other region in the world. Gas from its rotting garbage is siphoned from decades-old landfills to produce electricity. Its apricot pits and sugar beet scraps are burned to generate even more kilowatts. California cars are the world's cleanest, and getting cleaner. So are its power plants, print shops, furniture mills, dry cleaners, and bakeries. Even its paints produce little or no pollution.

On occasion, U.S. politicians find it convenient to invoke California, to say, for example, that the United States has the world's toughest air pollution rules. That's a true enough statement, but only if California is included, because the requirements established by Washington, D.C., have been weaker than those of Germany and Japan, not stronger. Indeed, for almost every weak rule laid down by Washington—whether it's pesticides in foods or smog in the air—California is the exception. And the meeting of October 1992 followed the pattern.

This conference on the business opportunities created by environmental concerns was organized by the South Coast Air Quality Management District, the government entity charged with regulating air pollution in Los Angeles. With a staff of roughly 980 and an annual budget of roughly $106 million, the district was easily the richest, most sophisticated and aggressive local air pollution control agency in the world.

Largely due to the efforts of the district and those of its sister agencies throughout the state, a wide range of new technologies was being brought to market, drawing an endless stream of foreign visitors from cities ranging from Tokyo to Berlin, all wanting to know what the cars and power plants of Germany and Japan will look like in another twenty years, after the rest of the world has caught up to California. Finally, Californians realized what had been obvious to outsiders: that the state was sitting on a mother lode of environmental innovation despite the national policy failures of the 1980s. The state's exports had already mounted to roughly $150 million annually by 1992, and visions of adding several zeros to that danced in the heads of the assembled manufacturers.[1]

There was some talk of devices that can be clamped onto smokestacks and tailpipes to eliminate pollution, but most attention focused on what the speaker from El Pollo Loco called "getting off the roof and into the kitchen," that is, eliminating pollution at its source. The specifics of El Pollo Loco's success story—how the chain eliminated smoke while cutting its labor, electricity, and other costs by one third to one half—fit the pattern of the California experience.

The stories, dozens of them, are the same. First the companies had attempted to continue business as usual and control pollution by weighing their roofs down with devices to capture, burn, neutralize, or otherwise treat the pollution they produced. But as demands for even greater reductions were imposed by the government, they began developing innovative new systems that not only made their plants cleaner and leaner, but spawned patented new technologies or approaches.

Consider, for example, Ray Turner of Hughes Aircraft. In the late 1980s, engineers throughout the world were turning to their space-age labs and high-tech computers to devise substitutes for chlorofluorocarbons, the ozone-destroying chemicals used for thousands of wildly diverse industrial applications ranging from air conditioning to plastic-foam egg cartons. At Hughes, CFCs were used to clean circuit boards, and the conventional wisdom was that while substitutes could be found for air conditioners and egg cartons fairly easily, the search for an alternative solvent that would meet the demands of the aerospace electronics industry would be long and tedious. For others it might have been, but not Turner. He mused over the problem during an afternoon's drive home, then went straight to his kitchen. There he came up with a substitute for CFCs that was not only ozone-safe, but cheap, plentiful, nontoxic, and nonpolluting. It was lemon juice.

Eighteen months of Hughes research has improved its citrus-based solvent, but the electronics cleaner being sold by Hughes today owes its origins to the lemon juice Ray Turner found in his kitchen that afternoon. Hughes markets it to electronics manufacturers throughout the world, pocketing not only those profits but another $4 million each year saved at its own operations because the citric acid substitute is cheaper than CFCs. All of which goes to show that there's money to be made in protecting the environment—lots of it.

In California, more than anyplace else in America, they know this. That's why the state leads the world in the following areas.

Turning sunshine into electricity. The world's largest solar thermal plant shimmers in the arid stillness at the edge of the Mohave Desert, turning out enough electricity to power a city the size of San Francisco and reducing air and other pollution accordingly. Elsewhere in the state, solar photovoltaic arrays on roadsides, mountaintops, and remote cabins free utilities of the need to string thousands of miles of unsightly (and costly) wires while again reducing pollution.

Harnessing the wind. Wind turbines generate more electricity in California than anywhere else in the world, except for Belgium. U.S. Windpower, one of only a handful of high-technology manufacturers of wind turbines in the world, is based in Livermore, California. U.S. Windpower ships its products throughout the world, generating not only kilowatts here and overseas, but jobs in California and elsewhere in the United States. Roughly 400 people work at U.S. Windpower's Livermore assembly plant, but other jobs are created in the Midwest, where the gears, generators, blades, and towers are made by companies ranging from Milwaukee Gear to Reliance Engineering.

Using waste energy. In most power plants, only about one third of the energy contained in oil, gas, or coal is actually used to generate electricity. The rest is vented to the air as waste heat. But not in California. There an aggressive program to encourage the utilization of this waste heat, a process called cogeneration, has resulted in the construction of hundreds of plants. Typically, some of the heat will be used to generate electricity while the rest is put to use running pulp and paper mills, refineries, sugar beet processing plants, food processing and canning lines, and a wide variety of other industrial facilities.[2]

Kicking the gasoline habit. Nowhere else in the world has a concerted effort to develop alternatives to gasoline yielded such a richly diverse fleet of cars, trucks, and buses as in California. The experimental, prototype, and production vehicles range from electric motorcycles to methanol-fueled water trucks, and almost everything in between. Confronted by the grim prospect that they could lose their markets to these alternative fuels, California's refineries responded to the threat by producing lower-polluting "reformulated" gasolines that have now gone on sale on street corners throughout America. In California, where these "environmental" gasolines originated, they often sell for *less* than the conventional, dirtier formulations.

The pursuit of zero pollution. Elsewhere in the world, the suggestion of eliminating pollution entirely is considered laughable. But not in California. Nonpolluting paints, barbecue starter fluids, cars, lawn mowers, stationary engines, and many more products are under development, at the verge of commercialization, or already on the market. Officials at the state and local levels not only don't laugh at the concept of eliminating air pollution altogether, they put their money on the barrel head, investing upwards of $10 million a year in state funds in technology development.

Saving money and avoiding nuclear. Some newly developed appliances, compact fluorescent light bulbs, for example, are so efficient that they save money for customers and utilities alike. Customers save because they use less energy, utilities because decreased demand means fewer new and expensive power plants must be built. California's energy conservation programs have been so spectacularly successful that they've saved consumers $1.9 billion in electricity costs and $1.1 billion in natural gas costs since 1977. In 1988, California was using 10 percent less electricity and 13 percent less natural gas than it would have without its conservation programs.

The state's Energy Commission predicts that by 1999 the combination of building and appliance efficiency standards, along with utility company conservation programs, will be conserving more than 11,000 megawatts of electricity per year—the equivalent of the output of twelve nuclear power plants.[3]

Shopping beyond borders. Just as the Germans and Japanese keep a sharp eye out for promising technologies outside their borders, so too do Californians. When air pollution officials at the South Coast Air Quality Management District heard of the Texas-based family business known as General Cryogenics (GCI), for example, they snapped up its technology. GCI had developed a system for mobile refrigeration trucks, like those that deliver ice cream and meats to supermarkets, that runs on liquified nitrogen and has only one moving part, a fan. Conventional systems—powered by noisy, dirty diesel engines, and leaking ozone-destroying CFCs through their hoses and fittings—destroy ozone and create smog, while GCI's system does neither. The district put a prototype on the road, and within months Thermo-King, the leading U.S. manufacturer of the old-fashioned systems, had bought up the production rights. Fuel cells being deployed in California are built in Canada, Japan, and Connecticut; zero-polluting paints are being developed in Michigan; cleaner-burning cars and trucks are being tested in Tennessee and Ohio. But all will eventually find a home in California.

In the 1970s, the state made several crucial decisions which led to the success of these programs. First, in response to the OPEC oil embargo, the state not only created the California Energy Commission (CEC), but charged it with protecting the environment.[4] In the Warren-Alquist Act, which together with its amendments and related statutes totals 241 printed pages, the legislature explicitly required utilities to "minimize cost to society . . . [and] improve the environment . . . through improvements in energy efficiency and development of renewable energy resources, such as wind, solar and geothermal energy."[5] Most state laws that regulate electric power companies focus exclusively on keeping rates as low as possible. In contrast, the Warren-Alquist Act required decisions to "include a value for any costs and benefits, including air quality." First passed in 1973, when it was vetoed by Governor Ronald Reagan, the bill was considered a second time in 1974—after the oil embargo—then signed into law (by Reagan), spurring interest in commercial development of solar, wind, geothermal, and other energy sources.[6]

The fledgling California Energy Commission proceeded to lay the groundwork for a switch to non-fossil fuels. Wind, solar, geothermal, and other resources were mapped. Studies were undertaken of the costs of alternative fuels and the barriers to their adoption. By 1992, the list of publications available from the Energy Commission was 39 pages long, ranging from a directory of certified lamp manufacturers to the *California Woodheat Handbook.* But it was left to a rival agency, the Public Utilities Commission, to take the step that would catapult California into the forefront of the alternative energy field.

The two agencies could hardly have been more different. The Energy Commission had been created in 1974, initially in response to projections that the state would have to fill its coastline with nuclear power plants to meet the electricity demand of its burgeoning population. The Public Utilities Commission (CPUC), in contrast, was a creature spawned by the Progressive movement. Its charter, established in 1911, was to ensure reasonable rates for customers of investor-owned utilities, which have a monopoly over their service areas.[7] The Energy Commission was headquartered in Sacramento, just a stone's throw from the nitty-gritty politics of the state capitol. The Utilities Commission was based in San Francisco, maintaining an aloof distance from politics. The Energy Commission was charged with implementing cutting-edge policies in areas like solar energy and conservation, while the Utilities Commission worked hand-in-glove with the state's powerful gas and electric companies to maintain the status quo.

Yet that was to change. When Reagan left office to launch his bid for the presidency, Jerry Brown became governor. Brown was widely ridiculed as "Governor Moonbeam" for (among other things) an energy policy described scathingly by critics as "windmills and woodchips." However laughable some of Brown's policies might have seemed, his support of tax credits, state purchases of renewable energy, and other incentives ultimately helped spawn the world's broadest and most successful energy diversity programs. Perhaps most importantly, Brown's appointees to the Utilities Commission, and their staffs, drafted and ultimately issued a rule that was to increase the amount of nontraditional energy in the state two-thousand-fold. To increase the state's use of renewable energy, the Utilities Commission had to devise a way to overcome what many considered the single biggest obstacle to the wide-scale deployment of solar, wind, and other forms of renewable energy—initial cost.

Much of the cost of generating electricity with coal, oil, or gas is in the fuel, which is paid and recouped by a utility in small increments over decades. In contrast, even though wind and solar generating systems rely on free, inexhaustible energy sources, comparatively large amounts of money have to be spent at the outset, then recovered over decades. Since solar, wind, and other renewable energy systems were unproven, utilities were unwilling to sign long-term contracts with alternative energy developers, and without those investors would not lend money to those same developers to build plants. To crack this vicious cycle, the Utilities Commission crafted Interim Standard Offer 4 (ISO-4).[8] It required utilities to sign long-term purchase agreements based on the price of generating electricity from oil, which was projected to continue rising through the decade.[9]

Able to show investors both customers for their electricity and a price likely to return a profit, solar, wind, and other companies were finally able to attract the capital they needed to construct bigger and better facilities.[10] With field experience, systems became cheaper and better still, and today wind turbines can generate electricity for less than a coal-fired plant and be built in a matter of months instead of years.

Ironically, the renewables industry was initially cautious about the new financial system, but by 1984 and 1985 the rush to sign with utilities had become a "veritable 'gold-rush' to sign contracts," in the words of one commentator.[11] The response was so overwhelming that the Utilities Commission suspended transactions under the new system on April 17, 1985, because by then "California was awash in new electric supply opportunities—and a potential excess of commitments." ISO-4 thus opened the door for the world's most robust solar, wind, geothermal, and cogeneration industries—all ready to compete with the existing utilities to supply the state's future needs for electric power.[12]

There are few success stories in the world that can rival California's spectacular increase in the number of kilowatts generated by something other than coal or oil. In 1979 there were only 5 megawatts of private power generation on-line in California. Ten years later there were over 9,000 megawatts. Moreover, the new plants were more reliable than the state's conventional power plants and provided electricity for an average price of 5.5¢ per kilowatt hour, one of the lowest prices in the nation.[13] Today, private generation in California represents approximately 35 percent of the private generation in the United States.[14]

Some technologies weren't ready for immediate deployment. They needed more work, something the new Energy Commission was able to pay for out of fines imposed on Exxon and other oil companies for illegal gasoline price gouging during the oil shocks. As the combination of incentives began to pay environmental and diversity dividends in the mid-1980s, another California agency, the Air Resources Board (CARB), escalated its attacks on the state's single largest polluter and energy user—the car.

In the 1960s, the government of California had become the first in the world to regulate air pollution from cars. When Congress enacted the federal Clean Air Act in 1970, the U.S. government established a national program. The new law prohibited the other forty-nine states from regulating tailpipe pollution from new cars, but allowed California to continue its policies. And it did.

In fact, much of the technology on today's cars was forced by aggressive California regulation. The state tightened emissions standards in 1975, prompting development of the first catalytic converters. It began phasing out leaded gasoline two years later—an action the U.S. Environmental Protection Agency failed to take until 1990, when the federal statute was finally amended by Congress to ban fuels containing the toxic metal. CARB regulated motorcycle emissions, tightened standards on diesel fuel, and required carmakers to manufacture pollution control systems that lasted longer and performed better in actual use.

In the 1980s California's regulatory bodies, ranging from the Air Resources Board to the South Coast Air Quality Management District, developed a new concept: treating the vehicle and its fuel as a single entity. The genius of this concept is that it opened a lucrative market to the rich and powerful electricity, natural gas, and alcohol industries, forcing gasoline—and the oil industry—to compete with other fuels for the first time since the abandonment of electric vehicles early in the twentieth century. Competition, in turn, inspired innovation, yielding a wide range of new and better fuels, engines, and pollution control devices.

It helped that the push toward zero pollution coincided with redoubled efforts at the Energy Commission to develop alternatives to oil. Both efforts, zero pollution and fuel diversity, can be attributed to the heads of the two agencies: Chuck Imbrecht at the CEC and Jananne Sharpless at CARB. The two had contrasting styles and different reasons, but shared a common goal. Imbrecht was dedicated to assuring California a degree of independence from oil because oil was imported, Sharpless because it was dirty. Together, their agencies launched a program to extend the development of alternative fuels to cars and trucks.

Several hundred miles to the south, Imbrecht and Sharpless found willing partners at the nation's premier air pollution control program. For years, cities and counties in the Los Angeles Basin had supposedly regulated air pollution. In reality, the rules had been weak and ineffective, at least in comparison to the task of cleaning up air so dirty that for one of every three days the pollution was a threat to human health.

In the mid-1980s, a new set of appointees took office in the Basin. Three of them—Norton Younglove, Larry Berg, and Hank Wedaa—stood out, each in his own way. Berg, a raspy-voiced Democrat known throughout California for his short temper and sharp tongue, formed the left flank and contrasted sharply with his two Republican colleagues. Younglove was a former high school teacher; Wedaa was, like many board members, semiretired. Of the three, Wedaa's position was the touchiest, because he represented Orange County, one of the wealthiest, most conservative areas of the United States.

Both heavy smokers, Berg and Younglove would often step outside board meetings for cigarettes and conversation. It was during such casual chats that the two often cut deals that later led to the nation's toughest air pollution rules. The irony that such rules were developed during cigarette breaks and by smokers was not lost on Berg, who fastidiously avoided exposing others to secondhand smoke. There was, he insisted, a sharp difference between hurting yourself and hurting others, a principle that was his guide during three terms on the board.

Dissimilar in virtually every other way, the three formed an environmental triad that launched some of the world's most stringent environmental measures: bans on aerosol containers and barbecue lighter fluids; mandates for car pools; and tight limits on pollution from dry cleaners, coffee roasters, print shops, and even restaurants. But perhaps their greatest contribution was to launch one of the world's most successful programs to develop antipollution technology through a new Technology Advancement Office (TAO), funded at a level that ranged from $6 million to $8 million per year. It was Berg's idea, but the other two quickly embraced it and it was speedily approved. Using money derived from pollution fines and a "dedicated" surcharge on car registrations, TAO scouted for promising technologies, recruiting companies to develop them when necessary, that could help solve the Los Angeles smog problem.

The South Coast Air Quality Management District's regulatory and technology programs worked together with the diversity and zero-pollution efforts of Imbrecht at CEC and Sharpless at CARB. Sometimes joined by officials at the Utilities Commission, the three agencies embarked on a technology development program unrivaled by those of the Japanese and Germans, yet at a fraction of the cost incurred by the U.S. government in the 1970s.

Their carrot-and-stick approach has yielded remarkable innovation.

GM's battery-powered car, the Impact, was developed in large part by the small California-based design team at AeroVironment, whose guiding genius is engineer Paul MacCready. The Impact accelerates from 0 to 60 mph in eight seconds, faster than even the best of its Japanese rivals. With a range of 125 miles on a single battery charge, many consider it to be the first truly practical electric car. GM has taken the car on a tour of Europe, where it is considered marketable because of the higher prices of gasoline there, yet without CARB's mandate for the sale of zero-emitting cars it likely would never have come into being.

Ford, Chrysler, General Motors, Nissan, Toyota, Volkswagen, and Volvo have all developed cars fueled by cleaner-burning methanol in response to California demands. Some of these models are now entering mass production and will be offered for sale in the next model year. Fearful of the market share it might lose to these newer, cleaner-burning fuels, ARCO mounted an effort to reformulate its gasolines, producing "environmental" blends that reduce some pollutants by up to 90 percent. Other refiners followed suit, and soon reformulation fever spread to diesel fuel as well. Today, cleaner-burning gasolines are being sold throughout America and the world.

Conventional gas-fired power plants produce up to about 200 parts per million of oxides of nitrogen, a leading cause of smog. They also operate at only about 32-percent efficiency, wasting over two thirds of their fuel. Faced with mandates to lower emissions levels, the makers and users of turbines began searching for ways to reduce their pollution. Aided by the California Energy Commission, they finally hit on the idea of using specially modified jet engines. Preliminary calculations indicate that these "chemically recuperated" turbines (essentially the same engines that fly Boeing 747s, but with some hydrogen mixed into the natural gas to make it burn cleaner and more completely) will operate at 60-percent efficiency and produce 1 part per million or less of the smog-causing oxides of nitrogen. The state's second largest utility, Pacific Gas and Electric, has asked for bids to build the machines, and if they work a global market will undoubtedly open up.

The primary purpose of California's aggressive programs is to reduce pollution, but governments and businesses alike are mindful of the money to be made in environmental protection. The U.S. Environmental Protection Agency, for example, estimates that the newly enacted 1990 Clean Air Act amendments, weak as they are, will generate up to $60 billion in new revenues and create as many as 60,000 new jobs over the next eight years.

This drive towards environmental innovation has been accelerated in California by the sharp and deep cuts in defense spending that have accompanied the collapse of the Soviet Union. Hughes is a prime example. The aerospace giant has historically derived the lion's share of its income from the Pentagon. In 1991, for example, 70 percent of the company's $7.8 billion income came from weapons work. Now, intent on boosting the non-defense share of its income to 50 percent of yearly earnings, Hughes sees the environment as a major market. Spearheading Hughes's entry into the environmental market is a newly formed 60-worker subsidiary, Hughes Environmental Systems. It specializes in cleaning up toxic waste sites and solving tough environmental problems—eliminating ozone-destroying chemicals, for example. Still another division at Hughes will soon begin mass-producing electronic controllers for General Motors' Impact, which is expected to go into production by the mid-1990s. The Torrance, California, plant is expected to employ 250 workers, all but one recruited from the staff that formerly designed radar for military jet fighters.

Hughes is not alone. Westinghouse has entered the environmental field as well, joining with Chrysler to develop zero-polluting electric vehicles—for sale in California. Northrop, producer of the $850-million-per-copy B-2 bomber, is also eyeing the possibility of entering California's electric vehicle and public transit market.

This move towards an environmental market is by no means limited to those who've survived on past military budgets. Nowhere is the result of large-company innovation more apparent than with automobiles. Vying for a share of California's market in 1998, when they must begin offering large numbers of zero-polluting cars for sale, the world's auto makers are unveiling an array of cleaner cars in addition to the models mentioned earlier. Ford touts a battery-powered minivan, while Chrysler boosts its "flex-fuel" Dodge Spirit, capable of running on either gasoline or methanol, or on a mixture of the two. In Munich, Germany, BMW has battery-powered models that promise competitive speed and acceleration. The campaign behind the Impact at General Motors (which, incidentally, owns Hughes Aircraft) is something of a turnabout for a company that has been a most strident opponent of environmental regulation for two decades. Instead of relying on electricity, the Japanese carmaker Mazda has designed its two-seat, 100-horsepower HR-X to run on hydrogen, which produces zero pollution when burned. Many consider hydrogen to be the fuel of the future, the first step on a path toward an entire economy—not just cars and trucks—that is zero-polluting.

One key to such a zero-pollution future is the development of devices to generate electricity from sunlight. Not surprisingly, California is already a step ahead, though it may require a trip to the mountains to see why.

About a hundred miles east of Los Angeles, hurricane-force winds whip the snow into six-foot drifts at the top of Onyx Peak, elevation 2,835 meters. The snow clutches at the steel and concrete bases of seventeen spindly radio towers and three squat, cymbal-shaped microwave relays that keep one portion of the San Bernardino Mountains linked with another. In 1989, the roar of diesel generators would have blanketed the peak. Today there is silence, for now the electricity is provided not by diesels, with their foul exhaust and deafening din, but by utterly silent, completely non-polluting solar photovoltaic cells which turn sunlight into kilowatts. More reliable than its coal, oil, and natural gas-fueled competition, solar energy is finally coming into its own as a way to make electricity.

California's Pacific Gas and Electric, serving the San Francisco Bay area, is widely regarded as the utility industry's most aggressive advocate of solar power, operating over 700 such installations. Many of these devices are built in California.[15] Siemens Solar in Camarillo, now owned by the German conglomerate, is still America's largest manufacturer, employing roughly 400 people.[16] Technological breakthroughs often occur in California as well. Recently, for example, Southern California Edison and Texas Instruments announced the development of a flexible, sandpaper-like PV system that uses minuscule silicon spheres embedded in aluminum foil (17,000 spheres per four square inches of foil) to generate electricity. Unlike the rigid PV panels in widespread use today, this "sun paper" can be made with cheaper, lower-grade silicon, thus lowering costs.

In an attempt to spur exports of California products, the Energy Commission gives grants to California-based firms so they can compete against firms from West Germany and Japan. While the amounts of money are admittedly puny—$50,000 per grant, at most—the efforts at least symbolize how California is thinking ahead on energy issues. An internal survey performed by the CEC indicates why: it revealed that 71 percent of those firms polled felt they could not keep up with technological advances occurring in the power plant business if they did not get involved in the international marketplace.[17] Recent grants include funding for the development of geothermal power on the Caribbean island of Montserrat by the Walnut Creek–based GeoProducts Corporation. Their $50,000 grant will help the firm complete a proposal for a 10-megawatt geothermal power plant that will provide all the electricity for the island.[18]

The CEC projects that future technology sales resulting from current state marketing efforts could top $1 billion.[19] One consequence of California's extraordinary growth in new fuels and technologies has been a new attitude among utility executives. The state's largest companies—Southern California Edison and Southern California Gas in the Los Angeles Basin, and Pacific Gas and Electric in the San Francisco Bay area—are now all numbered among the advocates of solar and wind power, as well as of fuel cells. The most outspoken official is undoubtedly David Freeman, former head of the Tennessee Valley Authority, who was selected in 1990 to become the general manager of the Sacramento Municipal Utility District (SMUD). "It should be the unwavering ambition of every utility executive in the country to put Exxon and Texaco out of business," says Freeman, and he means it.[20]

A garrulous man with a quick smile, Freeman wears a ten-gallon hat (a going-away gift from his previous job in Texas) everywhere. Since taking up the reins of the Sacramento Municipal Utility District, Freeman has placed the company at the forefront of the development of new ways of generating and using electricity. The utility has begun new initiatives to install fuel cells, operate electric vehicles, and build new solar generating capacity. Freeman is a particularly enthusiastic advocate of alternatives to coal, oil, and nuclear power because his experience has taught him that California's past successes and future promise could easily spread to the rest of the United States.

Before joining SMUD, Freeman had served as an energy advisor to presidents Nixon and Carter, overseen the landmark Ford Foundation Energy Project of the early 1970s, *A Time To Choose*, and headed two different utilities; the Tennessee Valley Authority and the Southern Colorado River Basin Authority. All of Freeman's experience has convinced him that the challenge confronting America can be overcome. "A nation that put an electric car on the moon," he is fond of quipping, "can surely put one on the road."

Indeed, the challenge to the United States of recapturing environmental and industrial leadership is no greater than dozens of others that Americans have confronted in the past three centuries. The nation's greatest advantages have always been the ingenuity of its people and its vast resources. The same is true today. Here are some facts to illustrate this point.

- Sunlight falling on the U.S. landmass delivers about 500 times as much energy as the United States consumes.
- The annual wind energy potential of just three states, North Dakota, Montana, and Wyoming, is equal to all of the electricity used in the United States in 1990.[21]
- The world's most advanced machines for exploiting solar and wind energy resources are, for the moment at least, still American.
- In 1985, more than 40 percent of the world's existing geothermal generating capacity was located in the United States; another 12 percent was in Mexico.[22]
- Exploitation of geothermal energy is growing throughout the world, and the United States is ideally positioned to take advantage of that growth.
- Because of nearly a century's experience in oil and gas operations, the United States retains a virtual monopoly in the time-consuming, expensive expertise required to discover and exploit geothermal resources.

Until assembly lines can be cranked up to churn out the tens of thousands of square miles of solar cells and the wind turbines necessary to exploit those sources of energy, the world will be pressing hard to maximize the efficiency with which coal, oil, and gas are used. Here, too, although there are rivals, the United States remains a leader. The smaller, more flexible aircraft-derivative turbines are made by only three major companies. Two of them, General Electric and Pratt and Whitney, are American. Of the world's first two assembly lines for manufacturing fuel cells, the Japanese own all of one and a large part of the second, but the majority of that second firm remains in U.S. hands, as does leadership in other fuel cell technologies.

The story is the same with respect to virtually every technology and every source of energy: unlike Japan, Germany, and its other economic rivals, the United States combines vast stores of the world's energy with a large fraction of its technological expertise. The nation's renewable resources dwarf the coal, oil, and gas in the ground, according to the U.S. Geological Survey (USGS). True, today's *proven* renewable reserves—energy which is identified and can be economically and legally extracted—are smaller, but we've not had a century of exploration and exploitation to develop them. Indeed, by almost every estimate, the energy that could physically be recovered from renewable resources almost certainly far exceeds current and foreseeable U.S. energy demand.[23]

Solar collectors alone, for example, in arrays covering less than 1 percent of U.S. territory—one-tenth the area devoted to agriculture—could make more energy than the United States consumes in a year.[24] Sunlight can be turned into electricity in two fundamentally different ways. In solar thermal systems, the sun's warmth is concentrated to superheat a fluid then used to produce steam, which, in turn, spins a turbine to generate electricity. In effect, the sun's fire is substituted for that produced in a conventional power plant when coal, oil, or gas are burned. The fire just happens to be 90 million miles away. In solar photovoltaic systems, the sun's radiation dislodges subatomic particles to create an electric current. Both solar thermal and PV systems work, though at the present state of development solar PV electricity costs two to four times as much as that of solar thermal systems.[25]

Those who question the workability of solar thermal systems need only travel to Daggett, on the edge of the Mohave Desert, where the world's largest system is located. As noted above, it produces enough electricity to supply the homes of a city the size of Atlanta or San Francisco. Price is not a big problem: the systems currently provide electricity at a cost of less than 10¢ per kilowatt-hour (kWh), and improvements in the technology could result in 30-percent cost reductions, making this technology cost-effective in virtually every U.S. and world market.[26]

Similarly, photovoltaic systems have been used for more than thirty years to power spacecraft, though today's major market is characterized more by millions of small, milliwatt-sized systems powering calculators and watches. Still, the only difference between these systems is size, and what works well enough to reliably power watches and calculators can do the same for offices, homes, and factories. Already the consumer market, which has been steady at about 5 MW per year, is expanding to larger systems for purposes like battery charging and walkway lighting. The largest new uses of photovoltaics today are for remote power for telecommunications sites like the one atop Onyx Peak, highway lighting and call boxes, navigation aids, security systems, water-supply pumping systems, remote monitoring, rural housing, and small villages.[27] At its present cost of 30¢ to 35¢ per kilowatt-hour, PV is too costly to compete against coal, oil, and natural gas for most uses. However, even at those prices, there is an untapped remote market of 200 to 300 MW in the United States and abroad. As costs drop—and virtually all analysts expect that they will—new markets will open up.[28]

Although solar-generated electricity may be the racehorse of renewable technologies, wind power is rapidly becoming the workhorse. Estimates of the amount of wind energy that could actually be recovered on U.S. territory, taking into account practical considerations such as the cost of acreage, also vary, but the consensus is that it is very large. One review of wind studies in the 1970s, for example, concluded that between 10 percent and 40 percent of U.S. electricity demand could realistically be supplied by wind power on land not currently used for other purposes.[29] The technologies needed to exploit these resources also exist, and in some cases are quite advanced.

Today, roughly 1.5 gigawatts of wind power are installed in the

United States, primarily in California. In 1989 alone these "wind farms" generated 2 billion kWh of electricity. Although the early development of wind power was closely tied to federal and state tax incentives, as well as to ISO-4 power purchases, turbines are now able to compete on their own, head-to-head with fossil-fired plants.

As is the case with other renewable technologies, wind power's early significant advances in this country have led to a worldwide technology development effort that far surpasses current domestic expenditures. In Europe, seven countries and the Commission of European Communities are each spending as much or more on wind energy research, development, and demonstration as the United States. A large Japanese manufacturer has also made a major commitment to penetrate the California market with the installation of hundreds of turbines.[30]

Despite the rapid gains that Japanese and German manufacturers have made in these and other technologies, the United States retains much of its advantage in expertise and experience. What seems to be lacking is leadership, for visionaries like David Freeman, Hank Wedaa, Chuck Imbrecht, Larry Berg, and Jan Sharpless are increasingly the exception, not the rule. Still, Berg likes to point out that the only difference between California and the rest of the United States is that there officials were willing to take a stand. Whatever the explanation, Washington would do well to study the successes of California.

PART III

THE NEW INDUSTRIES

Chapter Six

Wheels

> I am inaugurating a program to marshall both government and private research with the goal of producing an unconventionally powered, virtually pollution-free automobile within five years.
>
> —President Richard M. Nixon, February 10, 1970

Sitting inside the makeshift assembly plant in Camarillo, California, and even with a gaping hole where the windshield belonged and wires dangling from the vacant instrument panel, it was apparent that this car, the Impact, was destined to make history. Even though it was a product of America, widely regarded as a technological graveyard for automobiles, the Impact was sleek and sculpted, able to out-accelerate its best Japanese competitors. Powered by thirty-two lunchpail-sized batteries, rolling on advanced aerodynamic tires mounted on lightweight wheels, and designed so its driver surveys the road through heat-filtering glass, the car was clearly destined to capture the hearts and minds of drivers everywhere, from Berlin to Tokyo. It would, many agreed, reestablish its maker, General Motors, as a global center of vehicle innovation and help rebuild the American automobile industry.

Designed to meet a California requirement that 2 percent of all that state's new cars sales be "zero-emitting vehicles" (meaning they could produce no tailpipe pollution whatsoever) beginning with model year 1998, the Impact clearly bore the imprint of its principal designer, Paul MacCready. The head of a small California-based design firm, AeroVironment, MacCready had agreed to team with GM in the development of the unprecedented car. He brought impressive credentials to the task. MacCready had already designed a succession of world-famous vehicles, ranging from the Gossamer Albatross, a human-powered airplane that had crossed the English Channel, to the SunRaycer, a solar-powered car that had won a trans-Australia race sponsored by GM. For his accomplishments, MacCready had already been named "Engineer of the Century" by the American Society of Mechanical Engineers, notwithstanding the fact that the century still had ten years to run. Five of his vehicles were already in the Smithsonian Museum. The odds were that the Impact would become the sixth.

For its part, General Motors had contributed the resources which only the world's largest and richest auto manufacturer could bring to bear. GM, its suppliers, and its subsidiaries provided tires, wheels, windshields, motors, and a wide array of other parts for the car. Response to the Impact was wildly enthusiastic. Rave reviews quickly appeared in periodicals ranging from the *Washington Post* to the *Scientific American.*

That was January 1990. Nearly twenty-four months later, the General Motors board of directors announced that production plans for the Impact had been shelved indefinitely.[1] GM executives denied that the U.S. automotive giant was retreating from its plans to build electric vehicles, saying that postponing production of the Impact would make it possible for the Big Three to cooperate. The *Wall Street Journal*, however, observed that "GM's retreat from its promise to mass produce Impacts before 1998 also could make it easier for the Big Three to declare that California's timetable is unrealistic. Big Three officials almost certainly will seek additional electric-vehicle subsidies from the Clinton administration."[2]

A few months later, General Motors did exactly that, soliciting massive federal aid from the newly elected president. At the same time, its corporate lobbyists were asking federal officials to overturn the California zero-emission vehicle (ZEV) mandate.[3] Though little noticed in Washington, GM was also busily seeking to overturn ZEV programs in statehouses along the eastern seaboard from Maine to Virginia. Officials in these Northeastern states had announced plans to adopt the California tailpipe emissions regulations in an attempt to curb air pollution. Lobbyists for the Big Three teamed with oil refiners in the attempt to reverse these plans, asserting that the standards were "unnecessary and cost-ineffective."[4] Studies commissioned by this coalition claimed that in New Jersey alone the requirements would cost 35,000 jobs and $250 million in lost state and local tax revenue each year, starting in the year 2000.[5] In Maine, the same study concluded, adoption of the California rules would cost 5,600 jobs, a decline in personal income of $300 million statewide, and a $35-million reduction in tax revenue. The region as a whole, it concluded, would lose 298,000 jobs.[6]

Ironically, just as GM, Ford, and Chrysler turned their backs on the development of electric vehicles and redoubled their efforts to kill the ZEV program nationwide, Honda—the car company that had benefited the most from Detroit's earlier mistakes—began moving in precisely the opposite direction. Honda's president, Nobuhiko Kawamoto, announced that the company had reached "a turning point," and was abandoning Formula One Grand Prix racing in 1992 in favor of the development of electric vehicles. Kawamoto said that Honda would move most of the 100 engineers assigned to Formula One racing to the task of developing battery-powered cars. Honda planned to market such cars beginning in 1998, when the California mandate requiring them would take effect. Battery-powered Honda scooters would be offered in the 1993–94 model year. The engineers would be specifically assigned the task of increasing the mileage of electric vehicles and the performance of their batteries.[7]

For Honda, a company that had honed its technological skills on the racing circuit, the abandonment of Formula One contention was not a trivial decision. Indeed, Kawamoto is himself an enthusiast whose commitment to Honda's racing program was so unyielding that he once resolved to quit the carmaker in the 1960s when it briefly considered abandoning Formula One competition. Nevertheless, it was Kawamoto himself who personally ordered the change, saying, "We are facing a turning point, where we now have to consider improving our social and natural environment as well as winning motor races."[8] Honda was not alone, for at the 1992 Tokyo Motor Show, Japan's eleven carmakers displayed show cars featuring alternate-fuel engines, electric power, and hybrids of both.[9] "On the surface," said one reporter, "the move reflects pragmatic image-building on the carmakers' part. Deeper down, it shows a determination to be armed with the necessary technology when governments in many of Japan's huge export markets enact ever-tougher rules on exhaust emissions and fuel economy."[10]

For those familiar with the missteps that have plagued Detroit since the early 1970s, the contrast between the actions of the Big Three and those of the Japanese carmakers have an eerie familiarity. Once before, Detroit had been confronted by environmental and energy imperatives that provided the opportunity to expand and consolidate its base in the American market, then use that as a springboard toward increased international competitiveness, and once before Detroit blew it. Now, presented with the opportunity to steal a march on their Japanese and European competitors in the development of new, environmentally friendly energy technologies, Detroit seems on the verge of yet another self-inflicted grievous injury.

Whether electric vehicles will ever come to dominate the global market in the way that gasoline- and diesel-fueled versions now do is uncertain. Clearly, however, the world is moving in the direction of electric vehicles in terms of the pollution they produce—or, more accurately, don't produce. This trend envelops not only cars, but trucks and buses, even ships and locomotives. One of the few environmental fields where U.S. requirements can still be fairly said to be among the world's most stringent is the control of air pollution from motor vehicles. With the rest of the world clearly moving in the direction of tighter standards, and the United States (especially California) leading the way, the Big Three have an invaluable opportunity to develop cleaner technologies for the U.S. market that can later be exported to the rest of the world. This is precisely the methodology which has been successfully employed by the Japanese and Germans—and, for that matter, by the United States itself in an earlier era, when Detroit pioneered innovations ranging from the electric starter to the rearview mirror.

Instead of squarely confronting the irresistible trend towards cleaner, more fuel-efficient vehicles and recognizing the invaluable opportunity this trend presents, General Motors, Ford, and Chrysler are responding almost exactly the way they did to the environmental and energy imperatives of the 1970s, and in all likelihood the result will be the same.

Because motor vehicles are, in the aggregate, the source of more air pollution than any other human activity, the spread of emissions control programs throughout the world is inevitable, and has already begun.[11] Germany has proposed that the entire European Community, the largest car market in the world, adopt standards similar to California's by the year 2000.[12] Of course, as mentioned above, the eleven Northeastern states have also proposed adoption of the California program, but Detroit opposes that.

During the mid-1980s, Austria, the Netherlands, and the Federal Republic of Germany adopted innovative economic incentive approaches to encourage purchase of low-pollution vehicles. Australia, Canada, Finland, Austria, Norway, Sweden, and Switzerland all decided to adopt mandatory requirements.

Areas of rapid industrialization are now starting to experience surges in air pollution like those that struck the more developed nations starting in the 1950s and 1960s. Cities such as Mexico City, Delhi, Seoul, Singapore, Hong Kong, Sao Paulo, Manila, Santiago, Bangkok, Taipei, and Beijing (to cite just a few) are already so beset with air pollution that some (Bangkok, Mexico City, and Athens stand out) are already in crisis, with alarming levels of pollution a daily reality. As the vehicle population has grown and become increasingly concentrated in the world's growing number of megacities, the number of people subjected to unhealthy levels of pollution continues to rise.[13]

After years of delay, the European Economic Community has begun to move rapidly toward state-of-the-art pollution controls. As 1990 came to a close, the European Council of Environmental Ministers reached unanimous agreement to require all new models of light-duty vehicles to meet emission standards roughly equivalent to current U.S. levels by the model year 1992–93. Further, they voted to require the Commission to develop a proposal which, taking into account technical progress, will require a further reduction in limit values.[14] Even rapidly industrializing "developing" countries such as Brazil, Chile, Taiwan, Hong Kong, Mexico, Singapore, and South Korea have adopted stringent emissions regulations.

Although many nations are responding to the air pollution threat with attempts to develop public transit systems, these typically require massive public investment and long lead times. So, notwithstanding suggestions that the sun is setting on the automobile as the dominant form of personal transportation, most experts agree that our children and grandchildren are likely to still be driving them. Lester Hoel, a University of Virginia professor and former chair of the Transportation Research Board of the National Research Council, commented:

> There is no evidence to suggest that the automobile will not be the predominant form of transportation in 2020. This hardy little beast is characterized by flexibility for the user, permitting him to go wherever he wants and whenever he wants. It has a remarkable capacity to survive, changing its color, size and shape to conform to the current environment. Although the vehicle may be technologically unrecognizable as it adapts to the computer age, its essential features will remain.[15]

Still, the cars driven by our children will be—in fact, must be—cleaner and more fuel-efficient unless humanity resigns itself to choking fogs of air pollution in all of its major cities. Again, whether these cars emit zero air pollution or merely almost zero, they will nevertheless be cleaner than those of today.

In no other nation should the opportunity for developing and marketing technologies in response to these environmental and energy imperatives be brighter than in the United States. Its citizens, who comprise only 4 percent of the human population, own 33 percent of the world's cars and trucks.[16] While sales of non–U.S. manufacturers throughout the rest of the world are increasing, the United States is still home to three of the five largest auto makers. Unfortunately, despite its ready access to the world's largest market, both history and Detroit's current posture suggest that the U.S. industry is not poised to take advantage of the opportunities. To understand this, it is helpful to review the history of automobile innovation in the 1970s and 1980s.

Emissions, Fuel Economy, and Design

When, in the 1960s, the government of California became the first in the world to regulate tailpipe emissions of air pollution, other state governments began to emulate California, and pressure mounted for the Congress to enact nationally uniform legislation. Although Detroit approved of the notion of a single national standard, it wanted loose controls and therefore opposed the limits proposed in the Congress. Complaining to the chief Senate sponsor of the legislation, Edmund S. Muskie (D-Maine), the president of General Motors stated, "General Motors does not at this time know how to get production vehicles down to the emission levels that your bill would require for 1975 models. Accomplishment of these goals, as far as we know, simply is not technologically possible within the time frame required."[17]

Overriding such objections, Congress enacted tailpipe standards which, though postponed and modified slightly, went into effect in the mid- to late 1970s. The U.S. carmakers' response to mandatory fuel economy standards, standards proposed in the wake of the 1973–74 oil embargo and later extolled as "a major triumph of national energy policy," was similar.[18] U.S. carmakers resisted the fuel economy standards bitterly, saying such regulations would "outlaw full-size sedans and station wagons" (Chrysler), "require [only] subcompact vehicles" (Ford), and "restrict availability of 5- and 6-passenger cars regardless of consumer needs" (General Motors).[19]

Foreign carmakers, especially the Japanese, set about complying with the twin requirements, developing the world's most aerodynamically slick production cars, the first continuously variable transmissions, and the first wide-scale use of four valves per cylinder. After personally visiting the world's major manufacturers as part of a survey of the global car market, one expert concluded that "compared to their industrialized competitors who are at the forefront, American light vehicle manufacturers appear to be at the 'back of the pack' in the race to develop more efficient technologies. . . . Fewer technologies, for example, have been moved from the laboratory bench to testing in a prototype vehicle. . . . On the whole, the gap remains glaring with little promise of being closed any time soon."[20]

One specific technology, four valves per cylinder, is emblematic of the U.S. plight. As described by Douglas Cogan of the Investor Responsibility Research Center, "Multivalve engine technology was invented in the United States more than 80 years ago. Virtually every racing car built since 1912, in fact, has featured a double overhead cam, multivalve engine."[21] Yet multivalve engines were not introduced in passenger cars until the early 1980s, and it was Japanese automakers, not Detroit's Big Three, who made the move. Stiff taxes levied on large engines in Japan after the oil price shock of 1979 encouraged automakers there to find ways to give small engines higher performance.[22] Multivalve engines offer high performance and fuel economy, and have helped Japanese automakers take market share away from their American counterparts.

Multivalve engines are now standard in virtually all Japanese cars. Meanwhile, the only American-made multivalve engine in 1990 was the sixteen-valve Quad Four built by General Motors, which sold as a $660 option for selected models. Ford and Chrysler had no multivalve engines of their own, although they both imported such engines from Japan for some of their sportier models, such as the Taurus SHO and the Dodge Stealth R/T.[23]

It would be a mistake to attribute all of Detroit's misfortunes to resistance by the Big Three to past environmental and energy requirements. But neither can there be any doubt that such resistance has played a significant role in the steady decline of the U.S. industry, for many of the technologies developed in response to pollution and fuel economy rules have enhanced performance, durability, comfort, and other qualities that consumers shop for. Radial tires improve road handling, while fuel injection increases performance and engine durability, and aerodynamic designs enhance appearance. On-board computers not only improve performance but facilitate new innovations such as anti-skid brakes.

Zero-Emitting Vehicles: Follow California

The American auto industry's best chance for avoiding the same kinds of mistakes with respect to the next generation of cars is to embrace California's campaign against polluting vehicles. It is a campaign with precedents. Although the suggestion that the conventional internal combustion engine be eliminated altogether was criticized during the 1992 presidential campaign as environmental extremism, the first modern president to propose such a program was Richard M. Nixon, in a "Message to Congress" in 1970:

> Based on present trends, it is quite possible that by 1980 the increase in the sheer number of cars in densely populated areas will begin outrunning the technological limits of our capacity to reduce pollution from the internal combustion engine. I hope this will not happen. I hope the automobile industry's present determined effort to make the internal combustion engine sufficiently pollution-free succeeds. But if it does not, then unless motor vehicles with an alternative, low-pollution power source are available, vehicle-caused pollution will once again begin an inexorable increase.
>
> . . . I am inaugurating a program to marshall both government and private research with the goal of producing an unconventionally powered, virtually pollution-free automobile within five years.[24]

Nixon funded the program at about $10 million a year for five years, after which it was merged with other federal research and development programs and finally dropped from view, according to Graham Hagey, who worked for the Nixon initiative.[25]

California's famous call for zero-emission vehicles is merely one part of a complex matrix of increasingly tighter standards that are to be phased in with time. The first ZEVs must be on the road by 1998, when they must account for 2 percent of new car sales, or about one in every 650 cars. In 2003, 10 percent of new car sales, or one of every 50 cars, must be ZEVs.[26]

California's matrix of requirements has triggered a range of innovation. Since their enactment, every major carmaker in the world, ranging from BMW to General Motors, has developed a battery-powered vehicle, and so have some utilities. To aid the American carmakers, the U.S. government has given the U.S. advanced-battery consortium $8 million to develop low-weight, high-power batteries.

Cars and trucks fueled by natural gas number in the hundreds of thousands, with huge fleets in Italy, New Zealand, and the Soviet Union. But none of these vehicles was optimized to reduce tailpipe air pollution, with the result that many, perhaps even most, were worse polluters in some respects than their gasoline-fueled counterparts. With the advent of the California LEV (low-emission vehicles) matrix, however, carmakers and natural gas suppliers have been collaborating in the development of vehicles designed from the ground up to burn natural gas, and have done so with astounding results: after 50,000 miles, vehicles not only met the ULEV (ultra-low-emission vehicles) standard but were 96 percent below it.[27]

Since electric vehicles could decrease U.S. carbon dioxide emissions by roughly 25 percent, according to some analysts,[28] battery-powered cars would deliver doubly good news: local pollution such as smog and carbon monoxide would be reduced by eliminating tailpipes, and global warming pollutants like carbon dioxide would be curbed by substituting more efficient central power stations for conventional internal combustion engines. By the year 2010 or thereabouts, electric cars could have a firm grip on the market, assuming reasonable progress has been made in reducing their weight and recharging time while extending their range.

Although most attention has focused on batteries as the source of electricity, fuel cells are increasingly attracting attention. Powered by hydrogen, a fuel cell vehicle would also produce zero pollution, but offer greater range and power.[29] One fuel cell bus, built by Ballard Power Systems in Vancouver, British Columbia, was completed in 1993. It uses a proton exchange membrane (PEM) fuel cell, which relies on gaseous hydrogen and was originally developed by the U.S. space program, although the patents were allowed to expire and were acquired by Ballard.[30] Another fuel cell bus is to be built by the U.S. Department of Energy. The U.S. version will be powered by a Fuji Electric phosphoric acid fuel cell utilizing methanol as a fuel. These buses illustrate the intrinsic advantages of fuel cells over the current generation of batteries: greater range and power density (and hence greater ability to haul heavy loads), and short refueling or recharging times.

With the U.S. government willing to contribute substantial sums toward the development of electric vehicles, and state governments intent on encouraging sales through tax incentives and mandates,

Detroit's manufacturers are perfectly situated to establish a dominant position. Due largely to the pressure from California, each of the Big Three is in a position to rapidly develop cleaner cars with fuels ranging from electricity to natural gas. The Impact, for example, is a genuine product of automotive genius that almost all observers agree would sell virtually everywhere in the world (although, ironically, its prospects are bleakest in the United States because gasoline prices remain so low).

Ford is already marketing a "flex-fuel" Taurus, capable of running on either methanol or gasoline, and on April 14, 1992, the company announced that the 1993 models of the Ford Escort and Mercury Tracer, powered by a new 1.9-liter engine, "achieved a major milestone, four years ahead of schedule, making it the first auto maker to meet the first in a series of very stringent air quality standards adopted by California for the 1990s." As Ford further noted, "The engine in these vehicles was designed and is manufactured in the United States, using U.S. components."[31] Meanwhile, Chrysler has developed the battery-powered TEVan.

Few experts are more aware of the global trend toward less-polluting vehicles than Michael Walsh, former head of the motor vehicle control program at the U.S. Environmental Protection Agency and now an international consultant. Walsh's clients range from the governments of Sweden, Austria, and a number of other European nations to the World Bank, for which he provides assistance to developing nations beset by pollution, such as Mexico and the Philippines. Softspoken and affable, Walsh is known throughout the world for his command of the details of vehicular pollution control. For more than two decades he has found himself siding against Detroit when it comes to pollution control.

Testifying before a Senate subcommittee examining the future market for a "green" car, Walsh told members that "high growth in vehicle production will remain the norm, at least for the next several decades as countries with much lower per capita vehicle populations advance economically. The major threat to this growth is the serious environmental problems which they cause." Walsh sketched the movement of the world towards three types of cars: cars with lower emissions of conventional pollutants, such as those that form smog; cars that travel more miles per gallon of fuel, thus reducing global warming; and cars that produce no pollution whatsoever. "The future of the automobile industry," he concluded, "is dependent on its ability to build clean, efficient vehicles."[32]

In fact, the basic technologies for achieving the three goals of cleaner cars, higher mileage cars, and, ultimately, zero-polluting vehicles are relatively straightforward. They also are reasonably well known. The technologies fall into two broad categories, those that reduce the work the engine must do and those that improve the performance of the engine itself.

Engine load can be reduced by streamlining body shells, thus decreasing friction with the air. Similarly, higher-pressure, stiffer tires reduce road friction, while lightweight, high-strength materials can lessen weight. Better air conditioners and heat-filtering glass to keep interiors cooler also cut fuel consumption, as do other improved accessories. The Impact incorporates many of these features: narrow, high-pressure tires mounted on high-strength alloy rims; lightweight body materials and heat-filtering glass sculpted into an aerodynamically slick body. Engine efficiency can be boosted through a series of changes such as increasing the speed of combustion, raising compression, or optimizing air intake and exhaust. Some of these techniques or materials are incorporated in Honda's Civic VX, which has a combined highway and city mileage rating of fifty-one miles per gallon.

Defenders of Detroit argue that higher mileage standards will force weight reductions which will, in turn, lead to more highway fatalities. This argument intentionally confuses fuel economy with weight and weight with strength. Cars can be strong yet light; they can also be heavy yet fuel-efficient. The proof of this is that since 1980 average vehicle weight has remained almost constant, while fuel economy has increased by about 20 percent.[33] Manufacturers were able to improve fuel economy without reducing vehicle weight by relying on technological improvements in engines, transmissions, aerodynamics, and other systems. The potential for making further fuel economy improvements without reducing vehicle weight remains large.

At present, the only methods of building zero-polluting cars are to utilize either batteries or fuel cells, although it theoretically should also be possible to burn hydrogen.[34] Critics make much of the limited range of batteries, though the Impact has a range of 125 miles per charge despite its reliance on lead acid batteries that are considerably more advanced than those found under the hood for more than sixty years but nevertheless rely on the same chemicals. Other electric cars using these and other batteries have shorter ranges, but all have ranges that are within the average American daily commute of roughly 23 miles round-trip. Although the only cars built with fuel cells to date have been one-of-a-kind, Ballard's bus took only eighteen months to assemble from start to finish. None of this is to minimize the engineering obstacles raised by the task of building an assembly-line zero-polluting car capable of withstanding ten years and 100,000 miles of wear and tear. Some of these obstacles will no doubt prove formidable—but so, too, will the profits be for the companies and nations that tackle them successfully.

In the final analysis, what is really at issue in the debates over tighter fuel economy standards and more stringent pollution requirements is not whether cleaner, higher-mileage cars will be built, but who will make them—Americans, Japanese, or Europeans. Detroit has already hurt itself and the millions of Americans that have depended on the American carmakers for jobs and investment returns.

The U.S. car companies once made nine of every ten cars sold in the world. In terms of global sales, nine of the twenty largest auto companies in the world now are based in Japan. These companies account for 30 percent of world vehicle production and may soon eclipse production by the Big Three.[35] Even worse news is the decline in Detroit's share of the U.S. market. From 1981 to 1990, the Big Three's share of sales in the United States dropped from 70.8 percent to 62.5. If Big Three losses continue at this rate, in the year 2000 over one half of all cars sold in the United States will be made by somebody other than GM, Ford, or Chrysler.[36] The competitive pressure on General Motors, Ford, and Chrysler will mount because altogether the world's forty major auto companies have the capacity to build five cars for every four prospective car buyers. And because the United States offers the largest and most open market, 70 percent of this excess capacity finds its way onto the North American continent.[37]

In short, even if U.S. car companies brought flawless judgment and impeccable timing to bear, prospering in today's global market would be difficult. The Japanese are fierce and tenacious competitors. The Big Three have fallen so far, so fast that arresting the descent will be an arduous and exacting task. Recent announcements of collaboration, both among themselves and with the U.S. government, are encouraging signs if they signal a genuine change in attitude on the part of America's carmakers. With their talents combined, and with access to the space-age breakthroughs of the U.S. defense industry, the Big Three can unquestionably challenge both the Japanese and the Europeans. Yet to be seen, however, is whether Detroit's new rhetoric reflects a sincere departure from a regrettable series of previous errors.

Chapter Seven

Clean Power Technologies and Cleaner Fuels

We are on the threshold of change that could vault humanity into a new era of highly efficient, low-risk productivity, a sort of second Industrial Revolution, comparable in importance and impact to that of two centuries ago.

—WILLIAM MOOMAW, Tufts University

A hard three hours' drive east of Los Angeles lies what might have been the future of coal. The path is arrow-straight along Interstate 10, past the town of Ontario into the foothills, and finally to the edge of the Mohave Desert at Daggett. The plant's name, Cool Water, is drawn from the ranch that once occupied the site. Though its name suggests an oasis of coolness, the plant sprawls in the heat, silhouetted against the harsh whiteness of the sun-baked, dusty soil. Neither coolness nor water are in sight.

It was here in Daggett, California, that a handful of U.S. companies allied to create a new technology for burning coal, a technology with the mouth-filling name "integrated gasification combined cycle." Sometimes it's referred to as IGCC, but often it's called by the more appealing and pronounceable nickname "Cool Water." There was little actually new about IGCC. Rather, the partners in this endeavor cobbled together two technologies that were decades old. One of these is *gasification*, a system of converting coal into a gas; the other is a *combined cycle*, using two turbines in the system, rather than one. In the first turbine, the coal gases were burned to generate electricity. Then the excess heat (merely vented to the atmosphere in most power plants) was used to drive a steam turbine, producing even more kilowatts—all with less pollution than the usual methods of burning coal.

For the plant's demonstration run, coal was hauled to California from throughout the United States. Regardless of the coal's origin or type—high-sulfur or low-sulfur, Eastern or Western, high-ash or low-ash—the Cool Water plant burned it well, eliminating up to 99 percent of the coal's sulfur contamination. It worked so well, in fact, that one official report from the Department of Energy glowingly stated that Cool Water technology could potentially halve emissions of sulfur dioxide, the principal cause of acid rain, while holding electricity costs to levels "equivalent to or, in many cases, lower than the conventional options."[1]

Not surprisingly, the report predicted that IGCC could become "an integral part of the modernization of America's coal-fired power fleet. It could be a crucial option if utilities are to meet future consumer demands and remain financially healthy."[2]

> Repowering [with IGCC] would improve emission control, boost energy efficiencies, and enhance the cost effectiveness of coal-fired power generation. In a combined cycle configuration, it could significantly increase an existing plant's electricity generating capacity, perhaps by as much as 170 percent. This could reduce or perhaps eliminate the need for some power systems to build new baseload plants during the 1995–2010 time period.[3]

There was good reason for this optimism. The Cool Water plant's emissions were an order of magnitude cleaner than those of the best conventional coal-fired power plants. For sulfur dioxide, for example, a new power plant would be allowed to emit an average of about six tenths of a pound per million Btus. The Cool Water plant, burning the same coal, would emit only four one-hundredths of a pound. Emission reductions for oxides of nitrogen were just as extraordinary.[4] If it were built with better turbines and gasifiers, a

Cool Water–type plant would probably achieve roughly 45-percent efficiency, compared to the current average of about 34 percent.

The Cool Water plant's performance was important because coal is both the most abundant and the most polluting of humanity's common energy sources. A technology which could successfully curb coal-derived air pollution at reasonable cost would quickly find a ready market both in the United States and in the rest of the world. The United States' coal reserves, for example, would supply the nation's energy needs for several centuries, freeing it from its bondage to Persian oil. Moreover, since the utility sector emits about one third of all carbon dioxide produced by humans, and roughly 80 percent of that is from coal-fired power plants, a technology like IGCC could provide a bridge, allowing the world to continue using coal until cleaner fuels are developed.[5] It could have worked. The technology, after improvement, held the promise of 90- to 99-percent reductions in pollutants causing smog and acid rain, and 30- to 40-percent reductions in carbon dioxide, the chief cause of global warming.

Today, the Cool Water plant sits idle. It generates no electricity, even for demonstration purposes. It has passed into the hands of Texaco, which supplied the gasifier. The U.S. oil company intermittently seeks to find new uses for the Cool Water plant, such as burning sewage sludge hauled from Los Angeles, but always to no avail. Some reasons for IGCC's demise in California are peculiar to that state, and they include its success in developing alternative sources of energy. Never very enamored with burning coal to start with, California discovered that generating electricity with cleaner fuels, such as sunlight and wind power, could be as cheap as IGCC. Moreover, the fall in the price of oil caused by the success of the Reagan cheap oil strategy was accompanied in California by a drop in natural gas prices as well. Cool Water simply couldn't make electricity cheaply enough to compete.

Nationally, federal air pollution regulations might have forced adoption of Cool Water technology, but Reagan's policies of "regulatory ventilation" caused them to be shelved indefinitely. If Congress had responded to the threat of acid rain in the early or mid-1980s by toughening air pollution laws, these might have forced the spread of IGCC. But acid rain legislation was stalled in Congress for ten years by the delaying tactics of the coal and utility industries. Even when finally enacted, controls on the pollutants involved merely encouraged utilities to switch from coals high in sulfur content (sulfur is the major cause of acid rain) to others with somewhat less of the contaminant. High oil costs could have revived interest in IGCC, but the cheap oil strategy dropped real prices to their lowest levels in a half-century, making gasoline cheaper than milk or even bottled water.

In short, Cool Water, like so many other promising technologies, was a victim of the energy and environmental policies of President Reagan, combined with congressional inability to enact laws when needed. As a result, in the United States the Cool Water technology is essentially gathering dust. While two federally funded IGCC plants are under construction, the private sector in the United States has seemingly turned its back on the technology. Only one private IGCC plant is currently being built, a 260-megawatt facility at Delaware City, Delaware, by Texaco and Mission Energy to burn petroleum coke.[6] The two plants being paid for with tax dollars under the Clean Coal Technology Program won't be operational until the mid- to late 1990s. One of these federal plants, a 120-megawatt facility being built in Tallahassee, Florida, will provide a test bed for a foreign competitor, Lurgi. A German company, Lurgi has gasifiers operating in South Africa, as well as at another U.S. federal Clean Coal Technology venture, the Dakota Gasification Project near Beulah, North Dakota.[7]

But this is not the case outside the United States. When the German, Austrian, Dutch, and other governments toughened their air pollution laws in the 1980s, first in response to the threats of smog and acid rain and then to curb global warming, interest in technologies like IGCC soared. Using its version of gasification rather than Texaco's, the Dutch oil company Shell built a new and better Cool Water–type plant at the Deer Park manufacturing complex in Houston, Texas. Shell is scaling that demonstration up to 250 megawatts for another plant being built at Buggenum in the Netherlands.[8] Shell's results at its demonstration plant were so promising that Dutch plans to build a huge new conventional power station have been shelved to await evaluation of the new Shell IGCC.[9] At the Deer Park plant, efficiency reached about 43 percent, compared to roughly 38 percent for Cool Water.[10] Meanwhile, in Wakematsu, Japan, at the Electric Power Development Corporation, officials are busily perfecting an improved version of the technology, "hot gas" IGCC, in which operating temperatures are boosted to increase efficiency.[11]

Even if U.S. interest in newer technologies had been spurred by higher oil prices or tougher laws, it's possible that IGCC might have lost the contest for coal contracts to another technology, one called pressurized fluidized-bed combustion (PFBC). This technology can also be traced to American dollars and innovation, though at first glance PFBC seems to be Scandinavian in origin. PFBC is marketed by Asea Brown Boveri (ABB), a giant Swedish-German-Swiss conglomerate which may be the world's largest power engineering firm. First demonstrated in Malmo, Sweden, PFBC was scaled up to commercial size at the Vartan plant in Stockholm, as well as in Escatron, Spain.[12]

But appearances are deceiving. PFBC's inception dates to British efforts in 1968, and, according to ABB's director of marketing, Krishna Pillai, "the early work was primarily funded by the United States Department of Energy (and its predecessors) and the U.S. Environmental Protection Agency."[13] Other early demonstrations were conducted at the Argonne National Laboratory, starting in 1973, as well as by Exxon, Curtiss-Wright, and New York University.[14] The Malmo plant was financed in part by the American Electric Power Company (AEP), and the pivotal demonstration of PFBC, showing that it could operate using high-sulfur U.S. coals, was at AEP's Tidd power plant in Brilliant, Ohio—part of the federal Clean Coal Technology Program.[15] Thus, PFBC's development was paid for in substantial part by U.S. taxpayers and by the ratepayers of a Midwestern utility—AEP—that is not only among the country's largest, but also one of its most tenacious adversaries of air pollution control. Throughout the 1980s, AEP staged massive and effective campaigns to prevent a tightening of U.S. air pollution requirements at the very time when it knew they could have been met because of its own work with the 15-megawatt demonstration in Malmo starting in 1976.[16]

Across the Pacific lies another reminder of U.S. indifference. On the shore of Futsu Bay near Tokyo is a 2,000-megawatt power plant fueled with natural gas. It is the cleanest, most efficient power plant in the world, producing enough electricity for roughly a half-million homes. Futsu operates at 47-percent efficiency, compared to roughly 34 percent for most U.S. power plants. It emits only 10 parts per million of oxides of nitrogen, pollutants that create smog. No power plant turns more of its fuel into electricity, and no other power plant eliminates more pollution. Futsu's secrets are not the product of Japan's vaunted high-tech industries or its legendary globe-girdling conglomerates. They are the results of "Yankee ingenuity."

The key to Futsu's extraordinary efficiency is its combined-cycle turbines, manufactured in Greenville, South Carolina, by General Electric; the secret of its pollution control is an add-on system for eliminating oxides of nitrogen called selective catalytic reduction (SCR), whose catalyst was invented by Corning, Inc., in upstate New York. To see these GE combined-cycle turbines operating at such extraordinary efficiency requires a trip to Japan because U.S. utilities, benefiting from cheap natural gas and coal, have been unwilling to build such a plant. Similarly, Corning, unable to find a U.S. market for SCR technology, sold the patent to Mitsubishi.

These four technologies are by no means the only victims of U.S. neglect. As the 1970s drew to a close, the U.S. technological lead in the development of cleaner-burning fuels and technologies was commanding, seemingly insurmountable:

- Nuclear power had been all but invented in the United States, and virtually all of the so-called free world's reactors (originally developed to power U.S. nuclear submarines) were American-made. Only Sweden and the Soviet bloc boasted reactors of their own design.
- Solar, wind, and geothermal energy all had been brought to their most advanced stages of development by the Nixon, Ford, and Carter energy independence programs.
- Fuel cells—compact, quiet, super-efficient, and super-clean devices for converting fuel into electricity chemically rather than by burning it—had been advanced to the verge of com-

mercialization by the U.S. space program. Work on fuel cells had moved in two directions: toward using them in vehicles and, more promising at the time, toward using them to generate zero-polluting electricity.

- High-efficiency appliances, ranging from light bulbs to heat pumps, were in various stages of development in U.S. labs and factories. These products—light bulbs, for example—could do their jobs with only a fraction of energy required by conventional versions, and by saving energy they also reduced pollution, because the cleanest fuel is one that's never burned.

In short, although electricity was still being generated in essentially the same fashion as it had been for most of the twentieth century, there was ample evidence that the world stood at the brink of a technological revolution—one which would be led by the United States.

Today that revolution has started, but the United States is in the middle of the pack and falling to the rear.

The world's leading manufacturer of solar photovoltaic cells is now Japan, while Germany is in third place and virtually certain to climb to second in the near future. America's largest factory is now owned by a German conglomerate, Siemens, while the nation's third-largest facility is half-owned by the Japanese office-machines giant Canon.

Wind power has been deployed most extensively in Europe, where the most advanced turbines are also in operation. In the United States only one manufacturer of large turbines, U.S. Windpower, has managed to survive.

While U.S. manufacturers remain strong competitors in the race to commercialize fuel cells, one of the world's first semiautomated fuel cell assembly lines is at Fuji Electric, employing a technology bought from a U.S. firm. The other is in Connecticut, where Toshiba has a significant but undisclosed share of the control. The world's first fuel cell–powered bus is operating on the streets of Vancouver, Canada, powered by a fuel cell technology developed and abandoned by the United States.

Although American firms have developed improved nuclear reactors, the most promising ultra-safe versions are Swedish, Japanese, and German. The Swedish-German-Swiss conglomerate Asea Brown Boveri has developed an "inherently safe" nuclear reactor called PIUS, which it contends will withstand any upset short of a terrorist crash of a Boeing 747 into the reactor. Japan continues to work on the breeder reactor, with the goal of ultimately becoming independent of all fossil fuels.

High-efficiency lighting systems, many of them developed or fostered by U.S. federal spending, are selling in the United States by the millions, but the dominant manufacturers are Dutch, Japanese, and German.

This loss of U.S. dominance has occurred across almost all fuels and technologies, injuring large firms as well as small. One example is the combined-cycle systems that are an integral part of the Cool Water technology described above. General Electric was and is easily the dominant manufacturer of turbines in the United States and globally. Turbines used to generate electricity are fundamentally the same, although much, much larger, as those used to hurl aircraft through the air. GE, the world's largest manufacturer of jet engines, has been the beneficiary of billions upon billions of dollars spent developing newer and better jet engines for fighters and bombers, and the advances in aircraft turbines led to improvements in the much larger turbines used to generate electricity. Thus GE's historic dominance of the markets for the two sorts of turbines, those used for aircraft and those used for power plants, has been in large measure an offshoot of the national defense program. However, from 1979 to about 1987, GE was dealt a series of grievous blows that have enabled two competitors, Siemens and Asea Brown Boveri, to establish a foothold in the United States and to mount a challenge to GE's global leadership.

The worst of these setbacks began in the winter of 1976–77, one of the coldest on record in the United States. Temperatures dropped so low that the Chesapeake and Delaware bays froze solid for the first time in memory. This bitter cold sent demand for all fuels soaring, especially the demand for natural gas because it was used for heating so many homes. As supplies shrank, pipelines to businesses were cut off, forcing thousands out of work. Shutoffs to homes were threatened as well, sending fear rippling through communities in the East and Midwest.

These shortages coincided with the development of a proposed national energy strategy by the Carter administration. Indeed, there were some who believed that the shortages were exploited by the gas industry as a means of leveraging a decontrol of prices. Whatever the explanation for the crisis, it provided the political momentum for a two-edged national policy. On the one hand, under the Natural Gas Policy Act of 1978, to encourage greater exploration and production prices were decontrolled in phases, with total removal on January 1, 1985. On the other hand, the Fuel Use Act prohibited certain uses of natural gas, including its use in utility company boilers, on the grounds that it was too precious to merely burn.

Because it was against the law to construct a power plant that burned natural gas, orders for gas turbines dropped precipitously.[17] GE's orders for its large "Frame 7" turbine went from sixty-seven in one year to only five the next. The company "reached overseas," in the words of GE executive Eugene Zeltman; its business swung from 90-percent domestic and 10-percent international to exactly the opposite. By focusing on foreign markets, GE was able to hold total sales essentially level, thus enabling it to survive in hopes that the U.S. policy might be reversed.

Eventually the law was repealed, but what revived sales of combined-cycle systems initially was sleight of hand. Although it was against the law to build plants for the purpose of burning natural gas permanently, temporary permits would be granted to a utility that pledged that the turbine was merely the first step towards the ultimate construction of a coal-fired system such as IGCC or PFBC. As prices of oil and natural gas began to fall, utilities started placing orders for turbines, supposedly as first steps towards the eventual construction of IGCC systems, but actually for the purpose of circumventing federal law. This revival of U.S. turbine sales caught the eye of foreign manufacturers, especially Siemens and ABB, which had been forced to develop environmentally superior machines by the newly tightened pollution controls of western Europe. Siemens, for example, had developed a "dry, low-NO_x" burner for its line of combustion turbines, allowing them to achieve very low levels of emissions of oxides of nitrogen without SCR, the add-on control system employed by GE at the Futsu plant.[18] By the end of the 1980s, Siemens was offering its customers gas turbines with NO_x emissions at the unprecedented level of 9 parts per million—10 percent lower than Futsu's SCR system. Today, eighty-four of these are either in operation or on order.[19]

On the strength of such performance, backed by warranties and an aggressive marketing program, Siemens has already sold units in Delaware and Virginia, and seems to have an edge over GE in California. Now, with the market for gas turbines swelling due to falling fuel prices and tighter air pollution controls in many states, Siemens hopes to use the environmental performance of these recent orders as a wedge to boost its sales even further. By offering machines that it guarantees to be as reliable as GE's but cleaner, Siemens aims to enlarge its share of the U.S. market that has long been the almost exclusive preserve of General Electric. To keep one step ahead of its U.S. competitor, the German conglomerate is continuing to focus on improving the environmental performance of its combustion turbine systems. One such effort involves adding a fully fired cycle to the turbine operation. The full-firing process takes the exhaust from the gas turbine and recycles it into the combustion chamber along with new fuel and air. This increases the efficiency of the system by between 3 and 6 percent, which in turn reduces emissions of carbon dioxide.[20]

GE's Zeltman criticizes Siemens' turbine designs, saying that they lead to harmonic vibrations that can damage machines and lessen performance. (These turbine problems, he says, "don't show up until you've run them for a while.") He also expresses doubts that Siemens and ABB will be able to mount a sustained challenge to GE's leadership because "other companies don't have incorporated aircraft turbine business where this technology flows from," a reference to GE's continued dominance of the aircraft turbine market.[21]

Perhaps. But Siemens has begun manufacturing gas turbines for the U.S. market from an old Allis-Chalmers plant in Milwaukee, Wisconsin, while Asea Brown Boveri has a steadily expanding factory in Richmond, Virginia.[22] Both companies are relying on the environmental performance of their machines as a pivotal selling point in their struggle to expand their shares of the turbine market.

"This is part of Siemens' corporate culture," said Siemens Power Corporation vice president John Bobrowich. "Everything we do is environmentally concerned, and we put 10 cents on every dollar of revenues toward research and development."[23]

For its part, Siemens succeeded in capturing a two- or three-turbine sale at Delmarva Power and Light on the strength of low emissions, while it managed to remain competitive at a repowering project at San Diego Gas and Electric on the same basis.[24] ABB officials attribute their winning bid to provide steam turbines for the 650-megawatt Doswell, Virginia, plant to superior efficiency, which translates to lower pollution. ABB gas turbines have been sold for projects in Denver and Weld County, Colorado, as well as South Fond du Lac and Watertown, Wisconsin.[25]

Although corporate sales are a closely guarded secret within the turbine industry, almost all observers agree that General Electric's market share has decreased over the past several years, due in significant part to the superior environmental performance of competing machines. According to one survey, General Electric still enjoyed the lion's share of 1992 worldwide gas turbine orders with 9,900 megawatts of a total of 17,748, or 55.8 percent. GE's lead in sales of steam turbines was considerably smaller, though still commanding, with 39 percent of the 1992 market. Part of this can be explained by the proliferation of turbine manufacturers. Although a few major manufacturers dominate the market, the total number of steam turbine makers has climbed to eleven and of gas turbine makers to fifteen, roughly doubling in the last fifteen years.[26]

The export market for power generation equipment is one of the world's largest. Unfortunately, America's share is dropping, according to the U.S. Department of Energy:

> The competitiveness of U.S. suppliers has been declining over the past decade. Although accurate statistics are not easy to find, one can point to the fact that U.S. exports of power boilers have decreased from a high of $1.5 billion in 1980 to less than $250 million in 1988 and 1989 combined. Similarly, export sales of steam and gas turbine generator sets reached a low between 1986 and 1988. Between 1986 and 1989, U.S. exports of power generation equipment to developing

countries ranged between $2 billion and $3 billion; Japanese exports to these same countries exceeded $3.5 billion in the same period.[27]

Add-on pollution control technology, like SCR, will also help determine success. As the Global Environment Fund, a mutual fund specializing in environmental investments, concluded,

> In Europe and Japan, where tough acid rain pollutant restrictions were implemented during the 1980s, many coal-fired power plants are already equipped with a second generation of scrubber technologies that has generally improved on the original versions developed by American companies. In these countries, the waste disposal problem is generally reconciled by elaborate recycling programs which turn the waste gypsum mud into wallboard and concrete, and produce sulfuric acid. As a result of their improvements on older American technologies, and experience in large scrubber installations during the 1980s, it is likely that European and Japanese firms will be strong contenders, often with American partners, in the scrubber installation market.[28]

None of these technologies is currently prospering in the United States, for a number of reasons. As the government's cheap oil strategy has driven down crude oil prices, the costs of coal and natural gas have fallen as well. With fuel artificially cheap in the United States, there is less incentive to pursue energy-efficient technologies. Because pollution controls in America are much more relaxed than those in Western Europe and Japan, there is also little incentive to pursue these technologies for environmental reasons.[29] While it's true that Congress enacted amendments to the federal Clean Air Act in 1990 that were widely characterized as strengthening the law, almost all the emission limits imposed on factories and power plants can be met with relative ease through the simple expedient of switching to lower-sulfur coals. (Where some states are moving toward adopting more stringent emissions limits on their own, "they could tip the scale," in the words of one executive.)[30]

Finally, because of ideological opposition to government support of technological development, funding for most of these advances dissipated just as they were beginning to fulfill their promise. Private sector funding followed the federal plunge. In constant 1988 dollars, company-funded energy research and development fell 30 percent, from $3.46 billion in 1979 to $2.42 billion in 1987, the most recent year for which company data are available.[31] Constant-dollar company funds for conservation and renewable energy technologies declined 83 percent from $1.6 billion in 1979 to $284 million in 1987. Company funds for nuclear energy research and development fell by 66 percent over the same period, from $262 million to $88 million.[32]

There is little to suggest that these trends will be reversed in the near future. The continuing commitment to a cheap oil energy policy persists in undercutting the ability of the new technologies to compete on the basis of costs. The government spending philosophy that envisioned virtually no role for tax dollars in the development and commercialization of tomorrow's technologies has also been repudiated; but no coherent alternative policy has been formulated, much less enacted. Finally, although an environmental policy that consisted primarily of ignoring a wide range of burgeoning threats has been replaced by a willingness to acknowledge their reality, there appears to be a reluctance to concede that environmental policy must rely on more than corporate volunteerism.

It may be virtually impossible for the United States to recapture all of the lost momentum in this field, for regulations and funding were both slashed, orphaning many systems just as they were beginning to show commercial promise. Still, the race to bring these technologies to market is by no means over. The United States is more than just in the running; it stands an excellent chance of success.

Chapter Eight

The Inevitable Solution: Zero-Polluting Energy Sources

We always talk about the private sector as being too concerned with short term profits, but in the public sector it's a new ball game every October when the new fiscal year begins. You're simply unable to carry a long-term policy.

—James Caldwell

Continue skirting the edge of the Mohave desert after leaving the Cool Water plant and after about sixty miles a shimmering vision will begin rising from the horizon. Soon the vision becomes a phalanx of giant mirrored troughs, each twelve feet long and as tall as a small ranch house. As the sun moves, the computer-controlled mirrors pivot, focusing intently on the sun's brilliance. Running down each trough's center is a black steel tube. Sunlight reflected from the silver Mylar skin of each trough is focused on the tubes, superheating the synthetic oil within them to 735 degrees Fahrenheit. The oil flows to a low block building where it is converted to steam that drives a turbine which turns a generator that makes electricity—all with zero pollution.[1]

This is the heart of LUZ International, the world's largest solar thermal plant and the envy of the world. Here, on sunny days a torrent of zero-polluting electricity cascades through the power lines into the homes and offices of Los Angeles, enough energy to meet the residential needs of a city the size of Atlanta. On cloudy days, turbines fired with natural gas are turned on, but that's not often.

LUZ electricity sells for nearly as little as 8¢ a kilowatt hour, and its price would probably have dropped below the cost of electricity produced from coal by the mid-1990s. However, that won't happen, for in 1991 LUZ filed for bankruptcy. Under the supervision of the bankruptcy court, the solar generators continue to operate, still producing electricity, but the corporation is insolvent. LUZ was, almost all agree, a victim of on-again, off-again federal and state policies and the collapse in oil and gas prices during the 1980s, driven by the Reagan administration's cheap oil strategy.

"Solar facilities are categorized by high capital costs and very low operating costs," said Scott Sclar, executive director of the Solar Energy Industries Association. After over a decade of plowing profits back into improvements in its technology, LUZ was clearly on the way to cutting its costs to 5¢ per kilowatt hour. "But before the company could do it, fluctuations in U.S. tax and regulatory policies, as well as state property tax policies, bankrupted the world's leader in solar thermal electricity."[2]

"They had a fire that had nothing to do with the technology. They offended the labor unions and got in a property tax snit with the county," recalled James Caldwell, former president of ARCO Solar and now an international alternative energy consultant.[3] But these incidents were not the true cause of LUZ's demise, he said. Caldwell, like almost all other observers, blames that on two factors: the failure of the state and federal governments to adopt consistent tax credit policies, and the steady, steep decline in the price of natural gas.

LUZ, said Caldwell, had lived a hand-to-mouth existence for many years. "Every year they were playing 'You Bet Your Company,' every year it was a crisis. Every year they managed to pull it off. But one year they just couldn't do it and they went under."

What had provided the margin of survival for LUZ, enabling it to steadily improve its technology and thus lower costs, was the renewable energy business investment tax credit, which enabled investors to write off 15 percent of the money spent on solar, geothermal, and other alternative sources of energy. The tax credit was enacted in 1976, together with the much more widely known residential tax credit, which enabled homeowners to write off 40 percent of their investments in alternative energy. Although the residential credit expired in 1985 in the face of oil industry and Reagan administration opposition, the industrial credits were allowed to continue, but only on a year-to-year basis.

This raised special problems for LUZ, because it built its solar thermal units but didn't operate them. Instead, it would build a segment, then sell it to investors, plowing profits back into research and development. "They would have had to start building by April and have it on-line by December to get the tax credit," recalled Caldwell. "When it was completed and sold, they would pay back the loans. The negotiations would go right down to the wire."

All in all, it wasn't necessarily a bad system if the federal tax credits (and the state credits piggy-backed onto them) had been a permanent part of the Internal Revenue Code, but they weren't. Instead, because each extension lasted a year or less, LUZ was constantly lobbying Congress and the California Assembly to enact yet another extension. Although that usually happened, the constant uncertainty left LUZ unable to plan and at the mercy of both banks and buyers.

Still, if all LUZ had confronted had been the inconstancy of state and federal tax credits, it might have survived. But it faced a more formidable federal policy, namely the Reagan cheap oil strategy.

Through steady improvements in its technology, LUZ had been able to drive its own prices down from 25¢ a kilowatt hour in 1979 to about 8¢ at the time of its bankruptcy.[4] But natural gas prices fell faster than LUZ's costs, so its margins were always razor-thin. This heightened the importance of the yearly renewal of the tax credits to LUZ's survival, and thus its vulnerability to the whims of investors, bankers, and politicians.

To Caldwell—who once ran America's largest manufacturing company for solar photovoltaics, a firm now owned by Siemens—LUZ's demise raises fundamental questions about the manner in which the United States determines its energy and environmental policies and hence the nation's ability to compete in the global marketplace of the future. Many believe zero-polluting technologies ranging from wind turbines to solar photovoltaic cells will form the backbone of the world's industrial infrastructure in the twenty-first century. Some, wind turbines, for example, can already produce electricity as cheaply as it can be produced by burning coal, and others are coming closer.

In 1980, electricity accounted for about 10 percent of final energy demand in the world, with most analysts projecting that percentage to increase.[5] In the United States, which is roughly representative of most industrialized nations, electricity's share of the energy market has risen from 24.4 percent in 1970 to about 36 percent in 1989, with nearly 70 percent generated by carbon-containing coal, oil, or natural gas.[6] Pressure on these fuels is bound to increase because roughly one third of U.S. carbon dioxide emissions, a chief cause of global warming, come from electricity production, and this share may rise to 45 percent by the year 2015, barring some change.[7] Electric utilities are also the single largest source of oxides of nitrogen, which lead to the formation of acid rain, smog, and fine-particle pollution.[8]

And yet LUZ was by no means the only alternative energy company to be hurt by U.S. policies of the last fifteen years.

Solar Photovoltaics

Despite their promise, solar thermal systems are practical only in areas of bright and plentiful sunlight. However, solar photovoltaic cells generate electricity without the need for massive mirrors and turbine backup systems. Such solar electricity is, in the words of Paul Maycock, former head of the U.S. government's program, "the ultimate form of energy democracy."

Once the system is installed, the energy is free, plentiful, and utterly nonpolluting. Photovoltaic systems can be bought off the shelf and erected in a matter of hours.[9] They'll power everything from homes to offices. Panels can be bolted to the ground or a roof, or even molded into shapely curves. The devices seldom fail. When one panel does, the rest keep on working. They work in bright sun or under skies dimmed by clouds.[10] But solar photovoltaics do have their negative qualities: they don't work at night, and even in the brightest sunlight a kilowatt of solar-derived electricity can cost five times as much as one generated from coal.[11] (However, new breakthroughs are announced regularly, and one major manufacturer now says its solar panels can produce electricity for 10¢ a kilowatt hour, a rate well below those that prevail in many areas.)[12]

As noted previously, photovoltaics, like fuel cells and many other advanced technologies, were initially developed to meet the needs of the U.S. space program for a reliable source of electricity for American satellites. The leap from space to the earth's surface got its start when, under the sponsorship of the National Science Foundation, a meeting of scientists, engineers, and others active in the development of photovoltaics was convened at Cherry Hill, New Jersey, in 1973. This meeting has been called "a landmark in the movement to make photovoltaic practical for use on the earth." With the impetus supplied by the OPEC oil embargo, the Cherry Hill meeting laid the groundwork for the start of solar PV development in the United States. Interest in ground-based applications took hold after the oil embargo of 1973, then accelerated further after the 1978 oil shock.[13]

Aggressive pursuit of solar energy for ground applications in the United States began with enactment of the Photovoltaic Research, Development, and Demonstration Act of 1978. With this act, Congress committed the nation to a program to reach a cumulative production of solar PVs by the year 1988 of 4 million kilowatts—enough to power about a million homes—and to reduce the cost to between 6¢ and 12¢ per kilowatt hour, depending on the quality of the sunlight and the cost of capital.[14]

The law called for $1.2 billion in government spending, spread over ten years. It was, in the words of presidential adviser Ira Magaziner, "groundbreaking—the first time in recent history that government had ever planned to create a new, non-military industry."[15] The man named to head the program was Paul Maycock, a federal bureaucrat who had begun his career in the 1950s working with Bell Labs in the development of the first solar photovoltaic devices.

A year later, in 1979, when the second energy crisis hit, Maycock's plan was already under way. He'd gotten thirty private companies, twenty-five universities, and four government labs all working on photovoltaics.[16] Another year later, the first year Maycock's program was up to pace, the U.S. government was spending $150 million on solar energy.[17] In short order, the program became what many view as a model for the difficult task of coaxing a technology out of the laboratory and into the factories.

Maycock and his colleagues identified the obstacles to commercialization of solar PVs, then methodically set about overcoming them. The first task was to educate engineers, architects, and other professionals. Maycock established an education program with at least three public forums each year for professionals in the field. A basic university research system was established to train engineers and others. "We probably generated five or six thousand professionals," Maycock recalled. The program, he said, was also intended to support "solid theorists throughout the country. [If] you have a guy who wants to make his career in photovoltaic instead of transistors or integrated circuits, you have to have universities with the ability to support graduate students." Simultaneously, Maycock confronted the major obstacle to commercializing PVs—their lack of durability and reliability.[18]

From 1978 to 1981, the Department of Energy (DOE) spent roughly $15 million to identify and correct a wide range of technical faults.[19] One specific technique employed was to purchase 5-kilowatt "blocks" from each of the major manufacturers. These were tested by the Jet Propulsion Laboratory for exposure to water, stress, and outdoor conditions. Soon the program had developed testing standards and procedures that were published and voluntarily adopted by the industry. In addition, DOE provided production-line supervisors hired from the U.S. space program. These supervisors oversaw the operations of major manufacturers to assure that quality was built into the products as they came off the assembly line. Jim Caldwell later recalled that "there was one guy over at the Jet Propulsion Laboratory by the name of Ron Ross who ran a very, very nice quality assurance program that to this day is really the defining way to mark quality in photovoltaics. Somebody says if it passes JPL tests—which were probably eight, or nine, or ten years ago—that means something. That's the 'Good Housekeeping Seal of Approval.' "[20]

The purchases also served to boost production volume, thus lowering costs. Within three years, module prices dropped 75 percent, from $40 per peak watt to about $10. Government purchases of modules reached a high of 25 percent of production in 1980.[21] To boost production even further, Congress encouraged federal agencies to buy and use photovoltaic systems under the Federal Photovoltaic Utilization Program (FPUP). Nearly 4,000 separate projects were planned and funded which collectively generated about 750 kilowatts of peak power. Most were for small applications such as water pumping, communications, lighthouses, navigational aids, and power for rural national park and military installations.[22] In addition, the government initiated a program to demonstrate solar units under a system of Program Research and Demonstration Announcements (PRDAs). Hospitals, shopping centers, airports, high schools, and a wide variety of other facilities were supplied with PV systems at sites ranging from Ohio to Arizona. Many of these are still in operation.[23] Finally, solar photovoltaic systems were added to the classes of equipment eligible for the federal residential tax credit.

To stimulate overseas sales of U.S.–made PVs, the government entered into a series of multilateral and bilateral agreements with other nations for the purchase and deployment of solar systems. A 500-kilowatt system was installed to provide electricity to two villages near Riyadh, the capital of Saudi Arabia, at a cost of about $17 million. A 20-kilowatt experimental system was built in Italy, and an agreement reached to construct a 5-kilowatt system modeled on one installed at the Schuchuli Indian reservation in the United States.[24]

Then, literally overnight, the program was shelved. The Reagan administration, opposed to such government spending on the grounds that it interfered with "free market" decision making, cut spending to zero. Congress resisted, providing some money, but only a fraction of previous levels. By the mid-1980s, the United States had fallen behind Japan as the world's largest producer of PVs.

Although U.S. officials focused their efforts on the development of solar systems to provide electricity on larger scales, efforts in Japan took a different tack. There, officials concluded that the way to win an edge in the technology was to establish a broad base, providing a beachhead in a product before other nations could get firmly established. This philosophy had served Japan well in establishing its personal computer, office copier, and other industries.[25] Calculators, watches, and many other consumer products quickly began appearing on the market powered by tiny solar photovoltaic units which, in every way other than size, are identical to those that would be used at massive central power stations. "Americans laughed when we started development of solar cells for consumer products," recalled Dr. Yukinori Kuwano, head of Sanyo's development.[26]

This strategy—establishing a competitive wedge by finding a consumer product niche and using it to quickly build production volume, then steadily improving quality until you dominate the market—has worked very well for Japanese companies. Sony used it when it marketed the first transistor radio in 1956 (the same year, incidentally, that three Americans received the Nobel prize for the invention of the transistor). Honda did the same by first building motorcycles, then gradually "upscaling" into automobiles. Similarly, while products such as calculators provide only a thin profit margin, they help a company kick off a new technology, then build quality and production in rapid, small increments toward high-profit, big-dollar markets. The key is to sell products, for it is the resulting field experience that rapidly and sharply reduces costs, making it possible to enter, then capture, a succession of niche markets. First calculators; later, power plants.

This period in the struggle between the United States and Japan for what many believe will be the dominant form of generating electricity in the twenty-first century was examined in some detail by Ira Magaziner, then an international consultant. He concluded his review as follows: "Dr. Kuwano, head of Sanyo's solar program, will himself admit that the race is not over. He knows Japan's factories are making more solar cells than America's, but the real profits, he says, will be in industrial applications, not calculators. By the mid-1990s, power generation will be 70 percent of the market. It's unclear who will win most of it. But Kuwano says Sanyo is poised."[27]

As the development of solar photovoltaics has continued to languish in the United States, Japan has redoubled its efforts to make the technology a commercial reality.

By the year 2000, the Japanese government aims to have 250 megawatts of PV capacity, boosting that to 4,600 megawatts in 2010.[28] To promote installation of PV systems, the government has streamlined permit requirements. PV systems of fewer than 500 kilowatts, for example, are allowed to hook up without a permit, although a notice is required. For systems of 100 kilowatts or less, neither notice nor permit is necessary. For larger systems, revised guidelines will expedite hookups.[29]

Major new financial subsidies and other assistance are designed to stimulate private purchases of PVs. For commercial enterprises which install PV systems, 7 percent of the PV system cost is deducted from taxable income. Loans are provided to such a company with interest rates as low as 4.1 percent. "Model" plants may receive a cash subsidy equivalent to 50 percent of the installation cost. As previously mentioned, the government is setting up in 1992 a new institution to finance PV installations in public facilities (schools, public halls) for two thirds of the total cost.[30] To build demand outside Japan, the government plans to install four model plants in developing countries. Mongolia is expected to receive two such plants.[31]

Ironically, Japanese companies have also received indirect help from U.S. government funding. While Sanyo was focusing on the development of the consumer market, another giant Japanese conglomerate, Canon, was concentrating on commercial power production by purchasing a near-majority of control in United Solar Systems Corporation (USSC). In April 1993, USSC, a jointly-held subsidiary of Canon and Energy Conversion Devices (ECD) of Troy, Michigan, made an announcement hailed as "one of the most important events in the history of PV" by an industry trade journal—that USSC would build a $30-million plant in Newport News, with government aid from the state of Virginia. The plant was scheduled to employ 300 people and be in full operation by 1995.[32] Although the exact amount of the subsidy from Virginia was undisclosed, the state's program offers up to $22 million in grants. The USSC technology is based on development work at ECD that was heavily supported by federal funding. Among those joining in the announcement of the new plant was USSC's president, Shin-ichiro Nagashima.[33]

The Japanese-U.S. joint venture has also benefited from a three-year, $6.26-million contract from the U.S. Department of Energy, which enabled it to develop what one official called the "Holy Grail" of solar photovoltaics. It is a large-scale solar panel that can achieve a stable conversion efficiency of 10.2 percent, high enough for mass-produced units to generate electricity at homes and factories for as little as 10¢ per kilowatt hour, which is well below the rates that prevail in many areas of the nation.[34]

Fuel Cells

Many experts believe that solar photovoltaics will eventually make it possible for the world's drivers to do what they've dreamed of for nearly a century—run a car on water. What makes this prospect more than science fiction is water's potential as a source of hydrogen. Water can be split into hydrogen and oxygen simply by running an electric current through it.[35] The hydrogen can then be used to run all manner of machines, from cars and buses to submarines. One of the best means for harnessing hydrogen power would be fuel cells, compact, quiet, zero-polluting devices which, not unlike batteries, produce electricity through a chemical reaction rather than by burning the fuel in a mechanical engine.[36] Although they run best on pure hydrogen, other fuels that contain hydrogen—natural gas and methanol, for example—can also be used; pollution then climbs above zero, but not by much.

Since the late 1980s, the U.S. Department of Energy has been developing plans to build a transit bus powered by a fuel cell. As we briefly noted before, in early 1993, such a bus became a reality; fueled by hydrogen, it hit the street with considerable fanfare. When Russian premier Boris Yeltsin and U.S. President Bill Clinton met officially for the first time, this bus shuttled reporters. Unfortunately for America egos, that summit, and the bus, were in Vancouver, British Columbia, where a Canadian firm has built the world's newest fuel cell–powered, zero-polluting bus. The proton exchange membrane, or PEM, fuel cell used by Ballard Power Systems was originally developed by General Electric (and powered the Gemini earth-orbiting missions), but the patents were allowed to expire and Ballard "just picked them up," in the words of one officer.[37] Today, with the aggressive support of the governments of both Canada and British Columbia, Ballard has fielded a vehicle that may herald a breakthrough in transportation technology.

In Japan, Fuji Electric has begun the semiautomated manufacture of another type of fuel cell, based on phosphoric acid as an electrolyte and able to run on methanol or natural gas. Because these PAFC units produce less electricity per pound and per cubic foot than the PEM versions used by Ballard, they are considered less practical for cars, trucks, and buses; however, they may be well suited for homes, offices, hotels, apartments, and other places where both heat and electricity are needed.
originally developed for the U.S. space program, but sold to the Japanese firm in the 1980s. Asked how much money the National Aeronautics and Space Administration spent in development of the fuel cell, the former head of the program at the Department of Energy, Graham Hagey, replied, "They poured money into it—probably several hundred millions."[38] The development was, he said, fast paced and spread over five or six years, a "crash program" in which money was of secondary concern.

Eventually these programs bore fruit. Fuel cells powered both the Apollo and Gemini programs, then were further developed for use in the space shuttle as well, where they are still used.[39] Its NASA mission complete, fuel cell development was shelved by the U.S. government. "Nobody really picked up on the technology," recalled Hagey.

But as the U.S. government was phasing down fuel cell development, Bill Podolney of United Technologies recruited gas utilities to support the development of a 12.5-megawatt unit based on phosphoric acid cells. In 1975 he convinced the predecessor of the Department of Energy, the Energy Research and Development Agency, to support the development of a 4.5-megawatt unit to be built in New York City for Consolidated Edison. Soon other manufacturers joined. "With the government getting into the support of fuel cells, the rest of it came along," Hagey commented. In 1978, the conservation programs at DOE began supporting fuel cell development for transportation.

In the early days, budgets for fuel cells were running $20 million to $30 million per year, soon rising to $30–40 million, then to the present level of about $50 million. (From 1975 to 1992, support within the fossil fuel programs for stationary applications totaled about $800 million.)[40] When Reagan took office in 1981, he sought to cut fuel cell funding to the low single digits, $4 million to $6 million a year. But, unlike those for some other programs, these proposed cuts were consistently rejected by Congress. Still, it was in this period that Japanese interest in the U.S. fuel cell program, which started around 1979 (according to Hagey), intensified.

In rapid succession, Japanese firms acquired part or all of three vital fuel cell technologies. The Energy Research Corporation (ERC) "went to Sanyo and Mitsubishi and entered into agreements for molten carbonate [fuel cells]," said Graham Hagey; even today, Mitsubishi's "carbonate stacks look very much like ERC's—almost carbon copies." Fuji bought PAFC technology originally developed by the Englehard Corporation for the space program. Toshiba acquired a significant but undisclosed share of another phosphoric acid technology, this one developed by a division of United Technologies called International Fuel Cell Corporation (IFC). It is this loss that has especially pained many U.S. observers, because this was thought to be perhaps the most promising of the near-term fuel cell technologies. Indeed, ten units based on this technology have been installed in Southern California to provide heat and power to hotels, hospitals, and offices.

Japanese manufacturers have approached the fuel cell market with the single-minded intentions that their colleagues were focusing on the solar PV business. Fuji Electric has established a semiautomated assembly line for production of the PC25, a 50-kilowatt cell that emits remarkably small amounts of criteria pollutants and, by virtue of its efficiency of roughly 41 percent, will have comparably low emissions of carbon dioxide as well.[41] Fuji is continually improving the fuel cell: since 1984 its power density has doubled while volume and weight have dropped sharply. More than seventy 50- and 100-kilowatt cogeneration packages are planned for European demonstration projects.[42] Six units continue to operate at the Japanese Rokko Island Test facility near Osaka, where utilities and manufacturers collaborate to field-test equipment under government auspices.[43]

Simultaneously, Japan's government and industry are aiming toward the utility market, where the IFC-Toshiba technology is likely to play a major role. In 1984 and 1985, Japan's New Energy Development Organization (NEDO) constructed two 1-megawatt phosphoric acid fuel cells.[44] One was for use as a central power station, the other for dispersed generation. Operational testing of these units was concluded in 1988. A 4.5-megawatt PAFC was constructed in the early 1980s, operated by the Tokyo Electric Power Company (or TEPCO, the world's largest private power company, with about 26 million customers), then disassembled for inspection. Based on this experience, a similar 11-megawatt unit was placed in operation by TEPCO in 1990.[45] That 11-megawatt fuel cell bears a Toshiba label on its massive metal case, but its contents were made by IFC in the United States.

In 1986, NEDO began a five-year project to develop on-site PAFC systems for commercial use as well. This project includes two types of systems. The first is methanol-fueled and designed for use in parallel with diesel generators on remote islands. It is installed at the Tokashiki Power Station, operated by the Okinawa Electric Power Company. Utilizing technology developed by NEDO, the fuel cell has achieved an efficiency of 39.7 percent, measured by higher heating value. This level of efficiency is, according to NEDO, "among the highest in the world for an atmospheric type system of this capacity class."[46]

A second experimental 200-kilowatt fuel cell—an IFC-Toshiba unit—utilizes natural gas to supply both electricity and thermal energy to an Osaka hotel, the Plaza. Because the unit produces steam at a temperature of 170 degrees Centigrade, the thermal energy can be used for air conditioning as well as space and water heating. The unit has achieved an overall thermal efficiency level of 80 percent, while emitting only 4 parts per million of NO_x.[47] Based in large part on successful operation of the IFC-Toshiba unit at the Osaka Plaza, the 200-kilowatt package is now being marketed extensively for similar cogeneration applications. Ten are scheduled for installation in Southern California, and several have already begun operation.

To spur private purchases of fuel cells in Japan, the government will pay buyers a subsidy equal to one third of the capital cost of the cell.[48] In addition, the government has adopted a requirement that electric utilities purchase power from cogenerators.

Neither the United States nor its state governments provide incentives even approaching the plentiful and focused aid that has consciously been deployed by the Japanese government for the express purpose of coaxing the fuel cell industry from infancy to maturity. Indeed, most Americans—even officials who are supposedly responsible for the nation's industrial development and environmental protection—remain ignorant of the strides being made elsewhere and insensitive to the implications for the future.

Wind Power

Driving west through the hills of California toward San Francisco by way of Altamont Pass has the surreal feel of a time leap into some future when giant, gracefully elegant windmills provide all the power the world needs. They stand there now like giant storks, with their propellers slowly rotating in the nearly constant breeze, cranking out kilowatts by the thousands.

Altamont is the world's largest concentration of windmills, and a testament to how rapidly a technology can develop under the shelter of a protective government like California's. Operating in this climate, one company, U.S. Windpower, has grown in twenty years from a small and relatively obscure corporation to America's largest manufacturer of wind turbines. In the process, the state established itself as the unrivaled leader in exploitation of wind energy. Officials attribute this success almost wholly to the policies of California, for while U.S. Windpower and a few other manufacturers prospered there, those who chose to risk business in the other forty-nine states by and large perished.

Wind turbines—devices for converting wind into useful mechanical or electrical energy—are among the oldest sources of nonpolluting energy.[49] With recent improvements, they are also among the most efficient. When deployed in large arrays, modern state-of-the-art machines can now generate prodigious amounts of electricity at prices which compete with those charged by fossil-fired power plants.

After languishing during the late 1980s and early 1990s, however, wind power is undergoing a rebirth in a few areas of the United States. In Iowa, for example, plans are under way for a $200-million wind farm to be jointly operated by U.S. Windpower and an Iowa utility. It would generate 250 megawatts, about enough electricity for 50,000 average households.[50] Iowa development is spurred by a state goal of deploying 105 megawatts of renewable energy, coupled with a requirement that wind producers be paid 6.1¢ per kilowatt hour for their electricity.[51]

Interest is also picking up in Minnesota, where Northern States Power (NSP), the state's largest utility and one of the nation's most progressive, is under pressure from state leaders to develop the region's massive wind reserves. The Pacific Northwest is also eyeing wind power, and California may be on the verge of mandating another large purchase by its utilities.

But wind power industry officials remain troubled. "By the time it takes to build a market in this country," said a concerned Randall Swisher, "I'm afraid the Europeans will be in such a position that we just won't be in the game." Swisher, executive director of the American Wind Energy Association, was brooding over the ability of U.S. Wind power and the handful of other remaining U.S. manufacturers to compete against European companies that are growing by leaps and bounds thanks to generous and single-minded government programs designed to firmly establish the technology.[52] In 1992, for example, European governments spent $200 million on wind power, compared to $23.4 million in the United States. "No matter how smartly we might spend that money, it's not the same level of effort," Swisher stated.

In Burkburnett, Texas, those words might have held a special meaning. There, Carter Wind Systems, owner of what some consider to be one of the world's best-designed wind turbines, has also turned outside the United States for funding—naturally, to Europe.

Carter's partners are from the United Kingdom, where officials estimate that there is enough wind energy potential to supply the entire electrical needs of the island nation.[53] To spur development of wind and other forms of renewable energy, the British government has created the Non-Fossil Fuel Obligation program (NFFO). Started in 1989 under a Conservative, Margaret Thatcher, the program requires Britain's regional electric distribution utilities to buy a portion of their power from non-fossil sources. Although the relevant legislation was originally aimed at propping up the nation's ailing nuclear industry, the government has awarded 102 megawatts of capacity to wind power in 1990, with another 600 set aside for the year 2000.[54]

The United Kingdom's is but one of many European governments hoping to capitalize on the largely unexploited resource. In response, Europe is experiencing an explosion of new wind installations, while growth in the United States has generally leveled off after a long period of being sustained by California. The difference, according to Randall Swisher, is that Europe has a plan and the United States doesn't. "Commercial markets," he has said, "are the key to the [wind] industry's viability." While matters have improved substantially since the "incredibly depressing 1980s," the United States still has only a "technology development program, not a commercialization strategy." Swisher points to the absence of any national goals in the United States for deployment of wind turbines, compared to European targets of 4,000 and 8,000 megawatts for the years 2000 and 2005, respectively. Their goals are to be achieved through a variety of national programs.[55]

Under the Dutch government's Integral Plan Windenergie (IPW), capital cost subsidies brought roughly 65 megawatts of wind energy on line by 1991, with almost that much planned for the future. The plan was replaced in 1991 by the Toepassing Windenergie in Nederland (TWIN) program, whose goal is to install 250 megawatts by 1995 and 1,000 by the year 2000, with aid of just over a billion dollars in government support. Bonuses are offered by the environment ministry for low-noise turbines and for machines sited in specially approved, less environmentally sensitive areas. In addition, utilities must pay tariffs to wind turbine owners ranging from roughly 6.8¢ per kilowatt hour to about 10.6¢, depending on the province.[56]

A Danish wind energy incentive program providing a 30-percent subsidy for the capital costs was established in the early 1980s, then phased out as the costs of generating electricity declined. By the program's end in 1989, about 2,500 turbines with a total capacity of about 205 megawatts had been installed. Currently, wind turbine sales are exempt from the national value added tax (VAT). In addition, to help overcome the high up-front costs of wind turbines, the government joined with its domestic turbine manufacturers to create the Danish Wind Turbine Loan Guarantee program to underwrite the repayment of debt used to finance purchases.[57]

The Italian National Energy Plan established a goal of installing between 300 and 600 megawatts of wind capacity by the year 2000. A January 1991 law provides assistance for both initial and operating costs: up to 40 percent of installation expenses can be reimbursed by the government, while utilities are required to pay about 14¢ per kilowatt hour for the electricity.[58]

Germany requires electric utilities not only to buy power from windmill operators, but to pay them 90 percent of the average price charged homeowners, businesses, and other end-use customers.[59] In the United States the price paid would be much lower because the utility would be required to pay only the utility's "avoided cost," or the amount that it saves by buying the alternative energy. The avoided cost is usually the wholesale price of generating electricity with the cheapest available fuel, which is often the dirtiest of all fuels—coal.[60]

The obstacles raised by avoided cost reimbursement are merely one of several barriers that caused the U.S. wind industry to decline sharply during the 1980s, even though by the end of 1984 many thousands of wind machines were producing electric power in the United States.[61]

Wind turbines have been plagued by nagging problems: wide swings in wind speeds create roller-coaster surges and dips in electrical output and sometimes damaging transmission lines, and blades collect debris that hinders performance. But year after year, manufacturers have doggedly tackled these problems, solving one after another. Improvements in blade design alone have boosted efficiency by 25 percent, while new gear mechanisms and generators allow the machines to squeeze more electricity from gales and breezes alike. New turbines can now generate electricity for about 5¢ per kilowatt hour, equal to or lower than the price of coal-fired power.[62]

As the technology has improved, its use has expanded. Given the short time within which a wind farm can be deployed and operated (from one to two years, excluding wind data–gathering), growth under favorable circumstances could be extremely rapid. It is possible that the U.S. market potential for wind turbines could be as high as 21,000 megawatts of electricity for the decade 1990 to 2000.[63] Yet once again, despite its once formidable technological lead and vast resources, the United States is falling behind (though not in relation to Japan, where wind turbines are a lower priority because of the amount of scarce and expensive land they would occupy).

It is the tiny nation of Belgium, long a leader in the production of electricity from wind, that may soon become the world's number-one producer, followed by the United States. (However, if the wind power systems deployed in California were excluded from these totals, U.S. levels would sink to a par with those of Japan.)[64] Thus U.S. producers continue to struggle for domestic markets while those of other nations are being nurtured by tax incentives, financing, and utility and other subsidies. In the United States these advantages are directed toward the support of the oil, gas, and coal industries and toward nuclear technology, in which the United States was once the unrivaled world leader for good reason: it was invented here.

Nuclear Power

Leaving Tokyo, the train will rock gently for nearly two hours as it speeds northeast for roughly eighty miles, leaving first the downtown then even the suburbs behind. As the miles recede, the homes grow larger and farther apart, and when the train finally halts at the village of Tokai Mura only a handful of passengers disembark.

Although there's little visible industry in Tokai Mura, it is here that the Japanese have pinned their hopes for eventually achieving energy self-sufficiency. Perhaps it is Japan's utter dependence on the rest of the world for fuel that explains the national commitment to one of Tokai Mura's centerpieces, the small experimental breeder reactor called Joyo, "eternal flame."

The Joyo breeder has been running for several years, and by late 1993 a larger version, Monju, is scheduled to begin operating. Officials expect the demonstration to be followed by three progressively more powerful demonstration reactors, culminating in 1,500-megawatt-scale commercial plants in the 2010 to 2030 time frame. There are those who doubt that Japan's plans, however well laid, will ever come to pass. Still, few are willing to reject the possibility out of hand, for the Japanese have demonstrated an expertise in nuclear power that is unsurpassed by any other nation. Japan operates nuclear reactors that together with those of France are among the most reliable, economical, and accident-free in the world.

The two reactors operating in Tokai Mura are cases in point. The Tokai I reactor, which is of a British General Electric design known as the advanced Calder Hall type, is the oldest in Japan. This age does not equate with danger, however, because Tokai I relies on natural uranium as its fuel, which is much safer than the highly radioactive and enriched fuel used by modern reactors. One of those modern reactors is Tokai II, immediately adjacent to Tokai I. Tokai II is a typical boiling-water reactor of U.S. General Electric design. It produces 1,100 megawatts utilizing low-enriched uranium as its fuel. This output dwarfs that of the older reactor (which produces only 166 megawatts) because the older, bulkier Tokai I design is less efficient.

Despite their differences in type, size, fuel, and age, the Tokai reactors share two qualities—safety and reliability. In this respect, both are representative of the entire fleet of Japanese nuclear power plants, which, with 42 units, is the world's third largest. Only the United States, with 110 units, and France, with 56, have more plants. Moreover, while nuclear power is clearly in disfavor in the United States, Tokyo remains devoted to it, planning to nearly double the number of operating reactors by the year 2000.

The operating rate—the percentage of time a plant is actually producing electricity—of Japanese reactors is consistently among the world's best. Between 1987 and 1992, for example, of the seven nations with most nuclear generating capacity, Japan's operating rate was first in one year, second in three, third in one, and fourth in one. The United States, by contrast, was sixth during two years, fifth in three, and third in one. Only England's rate was consistently worse than that of the United States and only Sweden's was consistently better than Japan's.[65]

Japan's safety record is equally high-ranking, with less than one reported incident per reactor per year, even though the number of reactors has been increasing. Experts attribute Japan's operating and safety performance to careful construction, conservative operation, and a rigid commitment to maintenance which requires every plant to be shut down regularly for comprehensive maintenance and inspection.[66]

In the United States, where nuclear power was born, its prospects are bleak. For a wide variety of understandable reasons, the public and the utilities industry alike have soured on nuclear power. Construction times have doubled and plant performance has remained generally mediocre. Capital costs per kilowatt of installed capacity increased by a factor of four in real terms for plants completed between 1983 and 1987, as compared with a typical plant completed in 1971. Not surprisingly, no orders for new nuclear reactors have been placed since 1978, and the thirteen orders that were placed between 1975 and 1978 were canceled or deferred indefinitely. Few, if any, new orders are likely in the United States for the rest of this decade.[67]

Advocates of nuclear power believe widespread concerns over global warming may provide a window of opportunity for the nuclear industry. This might happen, for despite the fact that some view the technology with a deep and abiding mistrust, the fact remains that there are more than 400 reactors operating in thirty-two nations, reliably generating electricity. The industry is approaching the half-century mark with the number of severe accidents in the single digits, and many of those can be attributed to failures on the part of the operators, not the reactors or their designs.[68]

Twenty years ago (perhaps even ten) a turnaround in the prospects for atomic power would likely have provided the U.S. nuclear industry with a massive infusion of orders. Today, however, such orders would very likely go instead to firms in Japan and France for a simple reason: U.S. vendors haven't been building new plants, but the Japanese and French have. Moreover, while the U.S. industry has been battered by cost overruns, poor performance, and the near-disaster at Three Mile Island, the Japanese and French systems operate with near perfection. Once again, there is an explanation for this decline in U.S. fortunes—the relative laxness of U.S. environmental and safety requirements.

Both the Japanese and French reactors were originally of American design, but over time these fundamental designs have been modified so extensively that in many respects they have ceased to resemble the Westinghouse and General Electric plans from which they originated. One of the two principal U.S. manufacturers, Westinghouse, joined with the French government to create the French nuclear power company, Franatome, and in the mid-1970s the French government bought out Westinghouse and continued the process of improving on the original designs. Similarly, when the Japanese began to build their nuclear industry, General Electric and Westinghouse designs were used, but over time, Japanese firms took over more and more responsibility, eventually displacing the U.S. firms entirely.

In the United States, the Congress and presidents have responded to public concerns over nuclear power in either of two extremes: canceling programs, as happened with the breeder reactor, or shielding the industry, as with the Price-Anderson Act, which can bar or reduce legitimate claims by the victims of a catastrophic nuclear accident. These wildly opposing approaches have hindered the ability of the United States to remain abreast of Japan, France, and other nuclear nations.

In Japan, Germany, and Sweden, by contrast, the response to public concern over the safety of nuclear power plants has been to allay public concerns by making reactors safer and better through improved design, construction, operation, and maintenance. For example, when concerns arose over fuel reprocessing due to fears that plutonium might fall into the hands of terrorists, the United States merely scrubbed plans to build such a facility. Despite pressure from President Jimmy Carter, the Japanese persevered, and today, unlike the United States, Japan reprocesses spent fuel, thus reducing the threat to public health and safety.

Likewise, as public apprehension has increased in the United States following the incident at Three Mile Island and the near-meltdown at Chernobyl, the Price-Anderson Act continues to limit the liability of owners and operators of atomic power plants. The Japanese responded to public concern by establishing rigorous preventative maintenance schedules.

Efforts by the governments of Japan, Germany, and Sweden to calm public fears have not been entirely successful. Sweden's citizens, for example, decided in a national referendum to outlaw nuclear power after the turn of the century, and public protests regularly mar the industry's reputation in Japan. Still, public apprehensions, and in particular constructive governmental responses to them, have provided the impetus in those countries for efforts to make nuclear technologies safer and hence more marketable.

As noted above, Asea Brown Boveri, for example, has developed a PIUS reactor which it claims is capable of withstanding floods, fires, explosions, and virtually any other upset, including earthquakes. Similarly, ABB has developed a means of immobilizing and storing waste that it says will keep it safe for 10,000 years. Whether ABB's designs can actually fulfill such claims is uncertain. It is clear, however, that should nuclear power actually be revived by concerns over global warming, those seeking the safest possible means of coping with the technology's dangers are likely to turn first to Japanese, German, and Swedish companies. Had the United States put in place the kinds of environmental constraints on nuclear power that are manifestly justified, safer technologies would have been developed in America as well.

PART IV

THE FUTURE

Chapter Nine

Green Prophets

Better, faster, cleaner, cheaper.

—H. JEFFREY LEONARD,
Global Environment Fund

Taped to the rear window of Jeffrey Leonard's office was a child's drawing in soft pastels. Below it, stacked on their sides atop one another were the two-volume *Japanese Company Handbook* and the *Korean Company Handbook*, each of the three a different color: canary yellow, teal blue, and ivory white. A few feet to the right, two diplomas hung on the wall, one from Harvard University, the other from the London School of Economics. On one corner of Leonard's desk lay a folded copy of the *Financial Times*.

Behind the desk was Leonard himself, his dress and demeanor that of the quintessential American businessman. He wore a long-sleeved blue pinstripe shirt and gray trousers. A monogrammed silver belt buckle glinted as he half-rose to shake hands. He was, he said later, a man intent on becoming wealthy—if not immediately, then by the time his children entered college. Of this there was evidence, ranging from the investment portfolios and corporate brochures that littered his desk to the newspaper advertisements taped to the wall.

Only on close examination did it become evident that the portfolios, brochures, and advertisements had two things in common:

the environment and thirty-eight-year-old H. Jeffrey Leonard himself. Leonard is president of the Global Environment Fund (GEF), a rapidly growing investment fund founded on the conviction that the world's economic future is tinged with green, and that there's a fortune to be made by those who understand it. He is one of a small but growing number of individual businesspeople and large U.S. corporations who, despite national government inaction, are blazing a new trail of environmental investment—and reaping the rewards. Before examining what policies the nation ought to adopt, we'll consider these people and their companies, beginning with Leonard and the Global Environment Fund.

"Better, faster, cleaner, cheaper" is almost a mantra for GEF and for Leonard, who repeats the phrase at least a dozen times during an interview.[1] A more elaborate explanation of the Fund's investment philosophy is found in an article written by Leonard for *In Business* magazine. "The confluence of environmental concerns and economic constraints lies at the heart of a dynamic current of change that is gradually but inexorably going to reshape the very pillars of our economy—the transportation sector, the energy exploration and production industries, the electricity generating sector, and the chemical manufacturing industries."[2]

Founded in 1989 with assets of about $5 million, the Global Environment Fund is the brainchild of Leonard and John Earhart, both alumni of the World Wildlife Fund, a Washington, D.C.–based environmental organization.[3] There are a dozen or so other environmental funds, but GEF differs from them in one very important way. Other funds are content to invest in companies that clean pollution up; they specialize in add-on systems ranging from leachate collection systems for landfills to flue-gas desulfurization systems for power plants. GEF eschews after-the-fact cleanup and instead pursues technologies that avoid creating pollution in the first place.

Being clean, however, isn't enough. To warrant an investment from GEF, a technology must also be "faster, cheaper, better." Make no mistake about it, despite their environmental backgrounds, Leonard and Earhart are investors first and environmentalists second.

"We are not 'futuristic' investors, searching to identify revolutionary technologies to save the world," declares a GEF annual report. "We look for growth-oriented companies that demonstrate the ability to harness basic technological capabilities with managerial skills, financial capital and market acumen more than we look for companies focusing on the development of a single technology or scientific advancement."[4]

GEF's brief but successful history suggests that they have been doing something right. The amount of money that they manage has risen steadily, jumping from $5 million in late 1989 to $20 million in 1991, then to about $52 million in 1992, with a goal of reaching $100 million to $150 million by mid-1994. GEF's funds are provided by a mix of fifty-six institutional and individual limited partners. GEF investments, when weighted to take into account the performance of the various national markets throughout the world in which GEF operates, have grown faster than the market as a whole. GEF has also sharply outpaced other environmental funds. In 1992, for example, the dozen major environmental funds declined in value by an average of 3.2 percent, while GEF grew 2.3 percent. On the strength of this performance, GEF's investors have shifted more and more money into it.

Initially, GEF sunk substantial sums into companies like Ben & Jerry's, The Body Shop, and other firms catering to young, environmentally concerned consumers. "In the beginning we wanted some consumer product kinds of opportunities . . . and Ben & Jerry's was a small company that nobody had ever heard of. It's had explosive growth, extraordinary publicity—in fact, hype—and to us, the hype and so on began to say, How much more environmental innovation are we going to get here?"

"So we're out of that," Leonard explained, in favor of firms heavy on technology. "Previously it's been low-tech or consumer kind of products that have made money in the environmental area. It's now an ascendance of technology—applying advanced technologies—faster, better, cleaner, cheaper technologies to solve environmental and competitiveness problems."

GEF's annual statement of 1991 explains that it "look[s] for companies whose products and services eliminate pollution in the production process, minimize and reuse wastes, introduce cleaner raw materials or fuels, and promote energy efficient production techniques. The moving forces creating demand for such new goods and services are often a combination of changing consumer preferences, competitive economic pressures and government regulations which work all together to displace previous ways of doing business."

In support of his arguments, Leonard cites GEF investments in a range of little-known companies:

Novo Nordisk. A Norwegian firm with $1.2 billion in annual sales, Novo Nordisk is the world's largest producer of insulin, and it has used its position and expertise as a springboard into two new fields: bioenzymes to replace toxic and hazardous chemicals in food processing and the production of paper, detergents, and other products; and biopesticides to eliminate chemical toxins. "They are completely replacing harsh chemicals with bioenzymes," Leonard explained. "In many cases that eliminates heat and it eliminates chemicals from the production process. They're the world leader and that market is going to grow 30 to 40 percent a year for the next decade." He expects similarly spectacular growth in biopesticides. "Novo Nordisk has bought some of the best little companies here in the United States and they're going to be the world leader in the biological pesticide area."

Leonard described Novo Nordisk as "totally committed" to environmental protection in contrast to many companies that talk about a commitment then fail to follow through. "DuPont, Dow, Monsanto—they've all seen the light, they're all green. But they're giving with one hand and taking away with the other—they continue their existing practices, and then they have an innovative side," said Leonard, adding, "I can tell you that Novo Nordisk is a company that, top to bottom, is not green for the sake of being green. It has seen the market opportunities."

Leonard attributed Novo Nordisk's commitment to the aggressive environmental regulation in the Scandinavian countries. "The Scandinavian companies have lived in a situation where public awareness, regulations, and sheer scarcity of resources have forced them to think more environmentally."

Isco. "Monitoring often becomes the leading edge of an industry," said Leonard, citing Isco, a manufacturer of equipment to measure water pollution. In anticipation of impending regulation of stormwater runoff (the massively polluted water that cascades from city streets into sewers, then into rivers and streams during storms), cities have been buying large amounts of Isco's equipment. As a result, Isco's 1989 sales of $24 million jumped to $41 million in 1990, to $49 million in 1991, and to $60 million in 1992.

Measuring and Monitoring Services. Similarly, Measuring and Monitoring Services, a New Jersey–based company that makes equipment for measuring "negawatts"—the electricity that is saved by conservation devices such as high-efficiency light bulbs and super-efficient heating and cooling systems—has seen its revenues jump. Why? Because the utility industry is likely to spend $60 billion on conservation by the century's end, and the only way to be sure a dollar spent will result in a dollar conserved is with measurement devices like those made by MMS. "Now they're getting written into all the demand-side management projects," according to Leonard. "They doubled in revenue from last year to this, and they'll probably come close to tripling their revenue this year and the next. I'd say that over the next six years, you're going to see a 40 to 50 percent compounded growth rate."

Bandag. One of the most worrisome wastes generated by the vehicles that ply the roads of the world is their used tires. When burned they generate energy but also produce prodigious amounts of pollution. Tires can be pulverized, then used to "asphalt" roads, but that wastes a valuable raw material. Bandag, the world's largest recycler of used truck tires, recasts them for reuse. GEF bought stock in Bandag at below $30 a share. It doubled in value within twenty-four months.

Magma Power. A California-based geothermal power company, Magma's price per share rose sharply from $20 to $36 in just a few months.

Although many of GEF's investments are in U.S. firms, it roams the world seeking opportunities, partly because "the lack of regulatory initiatives for more than a decade left U.S. companies in a weak and technologically inferior position to many foreign competitors."[5]

GEF endeavors to invest in the "positive side of the industry rather than the regressive, remedial side," a policy that's proved successful. "Where we can get our greatest rate of return is by investing in things that are replacing old ways of doing things, eliminating, say, fossil fuel production or whatever it might be." Jeffrey Leonard foresees no limit to GEF's growth because the need for environmental protection, and its definition, "will always evolve." Still, he cautions, "Don't get me wrong that this is easy. The markets unfold erratically. We've seen zillions of great, wonderful, earth-saving, cleaner, faster, better, cheaper technologies never get developed."

One company that has eluded Leonard so far is Trigen Energy, based in White Plains, New York. As early as the mid-1970s, Trigen foresaw the coming demand for technologies that could simultaneously reduce both carbon dioxide and CFCs. One of the best of these is "tri-generation," the simultaneous generation of electricity, heat, and CFC-free chilling from a single source of power, such as a turbine. At the same time, district heating—the distribution of heat to homes, offices, or apartments from a central location—is common throughout Europe and can also be found in the United States at roughly 2,000 university campuses, many military bases, and some cities. Trigen established itself as a prime converter of new or existing district heating systems to tri-generation, winning contracts in Nassau County and in Kansas City, Tulsa, Trenton, Chicago, and a handful of other cities. In the process, its revenues have increased roughly 1,000 percent to over $60 million per year. A privately held company, Trigen has enjoyed such prosperity that it has no need for the capital that Leonard and his colleagues bring to other companies, leaving GEF almost yearning for the day when Trigen goes public. Until then, Trigen is the one that got away.

From one corner of Anne Shen Smith's U-shaped desk at the Southern California Gas Company, two of the forces that rule her world can be seen immediately. One is her three-and-a-half-year-old daughter, whose face smiles shyly from photos, eyeing visitors. The other looms outside the glass in Smith's eighteenth-floor office—Los Angeles smog.

The smog season is only beginning, but the brooding cloud of grayish-brown pollution is palpable. Without looking, Smith is aware of that smog, as are most of her fellow executives, because it could be either the instrument of their company's downfall or the source of its resurgence.

The struggle to rid California of the noxious cloud has yielded the world's most stringent air pollution regulations, so tough that they are compelling a redefinition of business in what is one of the most mature industries in the United States. One way or another, this redefinition will transform the Southern California Gas Company, America's largest gas utility. Either it will shrink in size and scope (and perhaps disappear altogether) or it will become a company providing a wide range of energy and environmental services, staving off new competitors and challenging old ones, including its archrival, Southern California Edison.

"Whether it is offensive or defensive," said chief executive Richard Farman, use of the environment as a business tool "is [a] key competitive strategy" for SoCalGas.[6] The company has struck off in dramatic new directions, developing natural gas–fueled car and truck engines that are 98 percent cleaner than the best gasoline-fueled versions on the road. It has also begun deployment of fuel cells. The company uses only recycled paper for its offices and recycled oil for its cars and trucks. Waste is minimized, according to official policy.

Recalling the time in the late 1980s when the company confronted emerging threats to its well-established monopoly, Farman said, "We consciously made a decision—and I will say I was initially pushed by people more progressive in their thinking at the time than perhaps I was—to look at the set of environmental issues in this way." Much of this rethinking was spurred by Anne Smith, vice president for environment and safety, who is charged with helping to develop a corporate strategy in which environmental protection serves the company as both shield and sword.

On the surface, the Southern California Gas Company does not appear to be a company especially in need of either a shield or a sword. In Los Angeles it's called simply "The Gas Company," as if it were the only one in the world. It serves one of every twelve natural gas customers in the United States. It has meters at 4.4 million homes, 250,000 small businesses, and another 1,000 large customers, ranging from cities to steel mills. It staffs its local offices with Spanish-speaking personnel, and its Multilingual Service Bureau has Cantonese, Korean, Mandarin, and Vietnamese translators. It is America's largest gas utility. SoCalGas's parent is Pacific Enterprise, a Los Angeles–based utility holding company with 1991 operating revenues of $6.5 billion.

Despite these impressive numbers, however, the one at the company's bottom line is disturbing to the corporate officers and shareholders: in 1991, Pacific Enterprises suffered an $88-million loss. One reason for this loss is that in the 1980s the federal government ended regulation of the natural gas industry, thus opening the way for companies from outside California to invade the territory. Disdainful of high-cost, low-return residential consumers, these corporate invaders have targeted the 1,000 commercial, industrial, electricity-generating, and wholesale customers to which SoCalGas sells about 75 percent of its gas.

Simultaneously, the company was confronted with the prospect of losing other customers because of air quality regulations. In an attempt to reduce the area's pervasive smog, local regulators decreed in 1989 that engines fueled with natural gas must add no more pollution to the region than comparable electric motors. The local air quality management plan was, Smith said, the "turning point." "The handwriting had been on the wall for a long, long time. But the view of environmental regulation had always been [that it was] an adjunct—it was a nuisance that was out there, but it was never really very serious. Very few people in the company really worried about it except the ones that were directly, functionally responsible for advocacy."

It was a measure of the low level of importance that the company assigned to environmental protection in those days that Smith and her colleagues were part of the public relations department. The dual shock of the "gas-on-gas" competition that came in the wake of federal deregulation and the advent of the proposed ban on fossil fuel combustion "jolted" the moribund company, said Smith. "There had always been an assumption in the past that there would be a future." Suddenly, there was "a realization that there may not be a future, or the future may look very, very different from what you were used to.

"There was a huge internal debate," she said, adding that "there are still some very hard feelings that exist today." The outcome of that debate was the conclusion that the key to the company's survival lay in not only responding to environmental issues, but using them as a springboard into new markets.

SoCalGas quickly became the first company to endorse the proposed air quality management plan to regulate air quality in the region. It consolidated its disparate environmental units to create a single Environment and Safety Department headed by one of its twelve vice presidents. Simultaneously, it embarked on a series of new technological initiatives:

Natural gas vehicles. Executives saw this as an opportunity to break into the most lucrative of all markets, transportation, because it provided the opportunity to sell not only vehicles, but also the fuel to run them. The company's development program for natural gas vehicles called for opening fifty-one refueling stations by the end of 1993, a pricing incentive to encourage drivers to buy natural gas instead of gasoline, and a suite of financial incentives for fleet owners to encourage the purchase of production-model natural gas vehicles.[7] In addition, it launched a program to produce car and truck engines "optimized" or designed and built to run on natural gas. Though millions of vehicles from Russia to New Zealand had been run on natural gas, none had ever been built with the specific objective of utilizing the fuel's unique qualities to produce lower pollution.

By the end of 1991, Ford had announced that it would introduce a fleet of about a hundred natural gas–powered light-duty pickup trucks in Southern California, to pave the way for mass production by the mid-1990s; it was producing the vehicles as part of a $23-million joint development program with SoCalGas and the gas industry's research organization, the Gas Research Institute.[8]

Fuel cells. Fuel cells can operate at twice the efficiency of conventional power plants and produce only a fraction of the air pollution. Officials saw fuel cells both as a means of enabling gas company customers to comply with stringent pollution controls and as a route to capturing electricity customers. Programs to bring fuel cells to the commercial stage were put on the fast track, and by May 1992, what was billed as "the first American commercial use of a fuel cell" began.[9]

The first unit was installed at the headquarters of the regional air pollution control agency, the South Coast Air Quality Management District, an ardent advocate of fuel cells. Others were installed at the Hyatt Hotel in Irvine, in Kaiser hospitals in Anaheim and Riverside, and at Kraft Foods in Buena Park, where they would provide not only electricity but also heat for warming air and water.[10]

Historically, electric and gas utilities have competed on a product-by-product basis, vying for sales on the basis of ranges, furnaces, and air conditioners. Fuel cells make it possible for a gas company to offer its customers electricity as well as a range of services and products. Farman expects the line between gas and electricity services to become "quite confused in the future," and he added that "for us to get more involved as an energy service company is going to come very, very easily."

Strategic alliances. During the same period, Farman assumed the presidency of the American Gas Association (AGA), the industry's trade association. Soon the AGA had forged an alliance with advocates of alternative energy and conservation, including the Solar Energy Industries Association and the Alliance to Save Energy. By April 1992, the three had published a study concluding that "a free market approach [and] an emphasis on efficiency and clean fuels can achieve significant reductions in emissions of carbon dioxide."[11] Soon afterwards, the company was reaching out to the environmental community, seeking to forge alliances with the mainstream organizations like the Sierra Club.

Low NO_x burners. At the same time, SoCalGas launched programs to preserve its markets by developing engines and burners that could meet stringent air pollution standards. Unable to find a U.S. manufacturer for such an engine, George Strang, vice president for engineering and operation support, sought out Japanese manufacturers. His quest was for a near-zero-pollution engine for use by local cities and towns in pumping well water. (Roughly 15 percent of these engines in the Los Angeles Basin were fired by natural gas, but users would be forced to switch to electricity unless Strang succeeded.) By mid-1993, the cleanest engine he'd found produced ten times the pollutants permitted by the Los Angeles standard, but Strang remained optimistic. "It looks like the Japanese system is probably our best chance, and we're going to ride that horse for another six months and see if we can make the right business arrangements. If we can't, then we'll be back on a technology scanning exercise again. Absent substitute technology, we will lose that market," an outcome Strang was determined to avoid.

A twenty-eight-year veteran of the company, Strang worked on corporate efforts in the 1980s to clean up toxic wastes and develop waste reduction programs, efforts that he considered a "watershed" in the company's attitude toward environmental protection. Those experiences proved to him that without the company's new, aggressive posture towards protecting the environment, "we would not have control over our own destiny and over our own costs. I think we'd [be] trying to meet the next regulation, and we'd be playing the role of the victim."

Sitting in the company's boardroom discussing this new attitude with Farman, it's difficult to believe that a major U.S. corporation would conclude that the environment is a strong enough driver that it could spell the difference between life and death for a company. But Farman says, "Yes, we believe it is such a driver."

Other companies may not yet realize that environmental concerns are forcing a redefinition of business. "I don't think it's all that obvious. It might be more obvious here in California because it is a combination of economic and regulatory forces at work, rather than just pure economic or environmental forces. We've got a combination of all three. In this business, as in any business, there will be some ahead of the curve and there will be some behind the curve. And those who are behind the curve get swallowed up, or go out of business, or are otherwise transformed—or don't survive."

In a pastoral, campus-like setting, Dave Chittick sits in a quiet corner office presiding over a small empire of engineers, lawyers, and public relations staff responsible for the worldwide environmental and safety performance of AT & T. This is no small task in a company that boasted $64.9 billion in 1992 revenues and $3.8 billion in profits. Long known as a homegrown U.S. company, AT & T is now truly a global enterprise, projecting that by the decade's end 50 percent of its sales will come from overseas. Already, 24.3 percent of revenues are from overseas goods and services.

With these profits come pitfalls as well, for, as a multinational company, AT & T must now concern itself not just with the environmental and safety regulations of the United States but with those of every nation in which it does business (almost all nations have such regulations).

It is AT & T's increasing exposure to the laws and mores of other nations that is fundamentally altering both the way it does business and the products themselves. For the first time, AT & T and its many business units are making product decisions, it says, on the basis of "the environmental soundness of a product's design, as well as its cost, functionality, and quality."

> There are many reasons to do so. First, customers worldwide are demanding responsible environmental performance and are putting their money behind their environmental concerns. Second, stringent regulations at all levels of government make end-of-pipe compliance ever more expensive. Third, environmental concerns are a strategic component of product globalization. AT & T Business Units must design products with global environmental regulations in mind or find their products barred or rejected by other countries.[12]

The challenges are awesome, with 340,000 employees at plants in Singapore, Korea, Mexico, and the Netherlands, as well as at twenty-one sites in the United States. In Dave Chittick's bookcase are copies of several books that provide some insight into the issues and strategies involved—*World Resources 1993*, *Sustainable Environmental Law*, and *Getting to Yes*.

Chittick's manner is every bit that of a vice president of a major global corporation: pinstriped shirt with cuff links and a white collar; a quiet but commanding presence. In conversation he notes that he is a Republican who supported George Bush and continues to harbor serious reservations about Clinton, yet this is an executive presiding over an environmental program that could, in many respects, have been designed by Greenpeace.

Chittick, who has been with the company for thirty-eight years and expects to retire soon, has been responsible for much of this change. On his bookcase sits a reminder of how much technology has improved in those thirty-eight years. It's a component from a switching device, roughly the size of a hardback book but so heavy that it must be held with two hands, a souvenir from his work roughly three decades earlier, when these components were combined to create a switching unit that was 107 feet long and 7 feet high and cost $1 million. Today, he reflects wryly, the equivalent function is performed by a device small enough to fit in one hand, at a cost of less than $5,000. He sighs at the prospect of missing the years of change that lie ahead, concluding almost mournfully that retirement is inevitable. "You have to do it sometime," he says, but clearly without relish for the prospect.[13]

Chittick has already presided over change that would shake most large corporations to their very core. Indeed, AT & T wasn't necessarily a willing, much less enthusiastic, participant early on. Confronted with the inescapable reality that the chlorofluorocarbons which it used in massive quantities were destroying the stratospheric ozone layer, AT & T knew that it had no choice but to identify and adopt substitute practices and products.

While the vast majority of other companies approached the task tentatively, AT & T foresaw the coming worldwide ban on CFCs and rushed to find its own substitutes for ozone-destroying chemicals, developing a compound based on a naturally occurring solvent found in cantaloupes. Another device, a "low-solids spray fluxer," eliminates CFC use in the manufacture of circuit boards. It's being used by thirty-two manufacturers throughout the world, roughly two thirds of which are non–AT & T companies.[14] These and other advances enabled AT & T to rapidly phase down CFCs, eliminating them at some plants well ahead of schedule. (The Singapore plant eliminated CFCs first, in 1990, and fourteen have since followed suit.)

The companywide effort to eliminate CFCs produced a pleasant surprise, according to Brad Allenby, research director for technology and environment of AT & T Worldwide Operations. "What the CFC episode did was force people to go back and explicitly review their processes. In doing so, they found that, most frequently, the processes could be made more efficient as well as switching the technology away from CFCs."

At roughly the same time, as AT & T began expanding into foreign markets, it started encountering increasingly stringent environmental laws in other areas. It was eventually forced to a corporate conclusion: increased, "ever-tightening" environmental regulations were "an element common to most nations of the world."[15] "So," explains one AT & T document, "the company's approach seeks to prevent pollution rather than rely on efforts that control emissions into water and air, but do nothing to prevent them."[16]

This new attitude has yielded some extraordinary results. In Richmond, Virginia, where the AT & T plant was emitting 4.6 million pounds of toxic air pollution in 1985, volumes were reduced slightly more than 1 million pounds by 1988, then to 500,000 pounds a year later, with a goal of eliminating them altogether by the end of 1993. In Atlanta, when the cost of hazardous waste disposal topped $1 million per year, the plant assembled a "regulated waste management improvement team" whose plan slashed wastes by 41 percent and disposal costs by $750,000 in a single year. In Columbus, Ohio, the AT & T plant not only reduced emissions of 1,1,1-trichloroethane, a toxic chemical, by 75 percent (with a complete phaseout slated for 1993), but reduced annual fuel costs by $60,000 as part of another program in which it burns methane gas from a nearby landfill to curb emissions of global warming gases.

This is not a corporation whose actions have grown out of a selfless desire to improve the quality of the world's environment. Both Chittick and Allenby grouse about the number and length of environmental laws and their apparent inconsistencies. It is clear that they, and AT & T, are tinged with green because it's good business.

Many of AT & T's changes have been small and incremental, setting the default mode of copying machines to the double-sided mode, for example, and switching to recycled paper. Many of these changes have resulted in savings: $26 million in reduced consumption of CFCs, for example, and another $4 million when AT & T shrank the size of its billing forms in order to decrease paper consumption. Now the changes are moving beyond AT & T's own operations to its products.

Allenby noted that customers throughout the world were demanding products that were environmentally responsible. He expressed personal skepticism that most consumers were, in fact, prepared to pay more for green products, while conceding that they express such a willingness in public opinion polls. But, he added, there are other kinds of consumers in the sophisticated large customer market—people like New York State, the Air Force, British Telecom, and various electronics companies. "More and more of the sophisticated customers are beginning to put environmental requirements into their RFP. That kind of thing can really drive corporate change."

AT & T is taking its environmental policy another step, laying plans to require that every new generation of products be environmentally preferable to the previous generation. "Now that's a *real* change in the way any company perceives its products and perceives the environment—because that's dramatically, radically different from the end-of-pipe approach," said Allenby. "What it reflects is a significant leap towards internalizing the environment and life-cycle environmental impacts of AT & T products into the design and operating costs." Reacting to a suggestion that such a policy might be based more on sentiment than business self-interest, Allenby responded somewhat angrily. "I don't think altruism has anything to do with it. If I were to go in and make an altruistic argument to a bunch of managers, I'd lose. There is, I think, an understanding that the environment is a competitiveness issue."

Chapter Ten

Facing the Future: Policy Recommendations

The best way to predict the future is to invent it.
—ALAN KAY, Apple Computer

The United States is confronted with a threat to its viability as a major industrial power in the coming century because it is failing to adapt to new environmental and technological imperatives. Some nations are responding to these imperatives themselves, others are reacting to the markets that the imperatives are creating. The United States is doing neither.

For a variety of reasons, the United States has neither a national technology commercialization program like Japan's nor the stringent environmental regulations, such as those enacted in Germany, that would stimulate new technologies by creating markets for them.

American politicians are fond of telling voters that American environmental laws are the world's most stringent, but it's simply untrue. The air pollution limits on power plants and factories in Japan, Germany, and most other industrialized nations are two to four times as stringent as those in the United States. So are recycling, environmental labeling, and pricing policies. Japan and Sweden tax pollution explicitly to discourage it. They and other nations also tax energy at up to ten times U.S. levels for a variety of reasons, including environmental ones.

To say that U.S. motor vehicle emissions standards are among the world's most stringent comes closer to the truth, though it's still shy of the mark. U.S. standards allow minivans, pickups, and other "light-duty trucks"—which are about 40 percent of sales—to emit up to 75 percent more pollution than cars. The federal requirements imposed on new cars sold in the United States are tougher than those in other nations, but they're still vastly weaker than the limits imposed by California, New York, and some other states, limits which, if the Germans and their allies have their way, will soon become law in the European Common Market as well. Perhaps more importantly, the United States continues to rely almost exclusively on cars and trucks to move its people and goods, while virtually all of the world's other industrialized nations, ranging from Brazil to Singapore, are shifting transport to cleaner, cheaper, faster, safer, and more comfortable alternatives.

U.S. technology development programs since the 1980s have lavished virtually all of their money on one specific fuel—coal—but have yet to commercialize even the Cool Water technology, once considering the world's most promising way of burning coal to generate electricity. Billions more have been spent on nuclear power, and those vast sums have failed to yield an inherently safe reactor. Recently, the Energy Policy Act of 1992 conferred a 1.5¢-per-kilowatt hour subsidy on wind, solar, and some other alternative fuels, but this hardly constitutes a strategy aimed at commercializing these technologies and then selling them here and abroad. It is, instead, merely another piece of the piecemeal approach that has characterized American policy since 1981, when hostility to environmental protection became the hallmark of American policy.

The environmental and technological leadership held by the United States in the 1960s and 1970s foundered on this hostility. Today, environmental regulations are too lax, energy prices too low, investment capital too expensive, and governmental assistance too limited to stimulate the commercial innovation that characterized those decades.

The first step toward restoring American competitiveness is to admit these harsh and unpleasant realities, and commit ourselves to solutions.

The beginning of any solution to these pressing technological challenges must be the expression of a strong policy commitment by our country's political leadership.

The president and the Congress might begin by fundamentally reforming both the rules that govern corporations and those that apply to their own elections, since the evidence suggests that America's current plight is due to some fundamental flaws in its institutions of government and business. Corporations should be explicitly required to include advancement of the public health and welfare among their objectives. Elected officials should be required to run for office the old-fashioned way, door-to-door and handshake-to-handshake, without benefit of any campaign contributions whatsoever.

But, leaving these large problems aside (solving them is not necessarily a prerequisite to confronting the more immediate challenge), once priority technologies can be defined, industry must then become a partner, helping to shape the necessary policies and programs to achieve economic and technological goals most efficiently and effectively. This may include cooperative research and development programs, cost sharing, and demonstrations. Representing a large consumer of energy and energy-using products, government procurement can be a test market and model.

The critical point is not to let debate about *means* interfere with agreement on *ends*. This is a lesson we can learn from both Japan and Germany. There, powerful agencies articulate standards and priorities, but they do not dictate how industry should meet them. Rather, they provide a framework in which industry and government can cooperate toward the attainment of national goals. German and Japanese companies play a much larger role than American ones do in shaping government policies to meet their needs in the most flexible and effective way, but unlike their U.S. counterparts, they do not employ literally tens of thousands of lobbyists and lawyers to engage in cynical, bitter, and time-consuming battles at every step. The question is not whether companies will eventually comply with new environmental and energy requirements, because in a global economy there will be no choice. Instead, the question is who will do so first, and therefore develop the products and processes that others will buy.

At the outset, a rational policy must acknowledge something that Americans seem to have forgotten: government is not a dirty word. Too often in the past dozen years, Americans have been told that they were faced with a hard choice between the heavy hand of government interference on the one hand and unfettered free enterprise on the other. However, government promotion of free enterprise has always been a central component of a wide range of U.S. laws and institutions. As the director of the Manufacturing Forum of the National Academy of Engineering and the National Academy of Sciences recalled,

> Government made major investments in transportation infrastructure to facilitate the exchange of raw materials and finished goods . . . invested in public education [so] every worker could read and do simple arithmetic . . . established a system of higher education to educate engineers . . . established laws to bring order to chaotic and antagonistic labor management relationships . . . adopted legislation and rules affecting banking and finance to enable the accumulation of large amounts of capital . . . [crafted the tax code] to provide incentives for investment in industrial plant, equipment, research and development . . . expand[ed] international markets for U.S. mass-produced goods by negotiating for reduction and removal of barriers to trade . . . [and] put in place a major system to support research on fundamental science, in the pursuit of major breakthroughs that might be exploited by industry.[1]

Canals were constructed by state and city governments in the early 1800s, opening up vast markets for commodities ranging from flour to salt pork. Railroads became the dominant form of transportation in the nineteenth century because of government intervention in the form of a long series of railway acts granting rights of way and in some cases an incentive bonus of title to every other section of adjacent land (these grants totaled roughly 155 million acres, an area about the size of the state of Texas). The automotive and trucking industries have, in their turn, enjoyed the government's generosity as well, through the construction of 3.9 million miles of streets and highways, including the 42,500-mile interstate system.

It is the government that protects the oil lifeline between the Middle East and the United States today, that over the last decade filled the Strategic Petroleum Reserve with a sixty-day supply of oil to guard against oil shocks, and that mandated the policies of conservation and fuel efficiency that cushioned the impact of Saddam Hussein's invasion of Kuwait. It was the U.S. government that took two laboratory curiosities, solar cells and fuel cells, and converted them to commercially marketable products. Finally, it was government that orphaned these technologies, abandoning them to America's economic rivals.

There is a limit to what individuals can do, whether they're persons or companies. No single corporation or ordinary citizen can rescue the United States from its current predicament. It is government, in the persons of our elected leaders, which must now point the way out. Assuming that government remains capable of acting decisively, what might a rational program resemble?

An Economic, Environmental, and Energy Policy for the 1990s

Books on national policy usually conclude with authors' lists of specific suggestions, often with a timetable. We also present specific options, but principally to make the point that there are literally hundreds of choices available. There is no single perfect solution, or combination of them. The examples of Germany, Japan, and California illustrate this point: each chose a suite of options suited to its unique character and resources.

When President Kennedy committed the United States to the goal of landing on the moon within a decade, he built a political consensus in support of that goal, then set about achieving it with an open mind as to which of the many options were the best. The result of that commitment to a "great new American enterprise"[2] has been a profusion of technologies followed by unquestioned and unrivaled U.S. dominance of the aerospace industry.

Despite this experience, there is a false perception, perpetuated by the Reagan and Bush administrations, that government cannot effectively promote technological innovation, a perception that has had many disastrous consequences. One of the more subtle is the absence of policy experimentation and the consequent narrowing of policy choices. With the exception of the emissions trading program added to the Clean Air Act in 1990 (a concept first proposed during the Carter administration, and a questionable one at that), national energy and environmental policy debates fixate on the same limited options that dominated public discourse more than a decade ago. Policy goals—cleaning the air, reducing oil imports, and so on—are overshadowed by a predictable exchange of heated opinions about the means of achieving them.

One of the best (or worst) examples is the debate over automobile fuel economy. The strategic importance of significant improvements in the fuel economy of gasoline-powered vehicles is indisputable. In all parts of the world, the compelling attraction of personal mobility has meant a steady growth in the demand for cars. With slight interruptions, the same has been true for vehicle usage. Barring radical improvements in vehicle efficiency and large-scale adoption of alternative fuels, the combination of growing populations and rising incomes will inevitably lead to an enormous increase in demand for oil and a growing dependence on Middle Eastern suppliers.

Rather than face this problem directly, Congress has haggled over two basic policy choices, higher corporate average fuel economy (CAFE) standards and gasoline taxes. The oil industry advocates a third strategy, more drilling, ignoring the environmental consequences and the inevitable continued decline in U.S. production no matter how much drilling is permitted. Each of these options is bitterly opposed by political interests able to block any action.

The automobile industry, behind powerful allies, has so far prevented any increase in fuel economy standards. Auto industry officials contend that CAFE standards disadvantage American car companies because the latter manufacture a high percentage of larger, less fuel-efficient vehicles. Auto industry officials argue that higher gasoline taxes are the best way to promote efficiency since, they contend, consumers will not pay the added cost of the necessary technology without the greater economic incentives. However, gasoline taxes—when they are enacted, which isn't often—inevitably fall far short of the levels required to shift behavior, and are usually offset by drops in the price of crude oil due to the continued success of the cheap oil strategy.

Moreover, while there is evidence that higher prices (achieved through higher taxes) are linked to higher-mileage cars, there is a limit to which elected leaders can or should raise gasoline tariffs.[3] It is worth noting that numerous polls show that while a near-majority of Americans are willing to pay 25 to 50 cents per gallon more in gasoline taxes for the sake of protecting the environment, many more of them believe that the proper solution is to mandate higher-mileage cars, even if that means higher new-car purchase prices.

The consequence of rejecting both gas taxes and higher CAFE standards is a vacuum in federal policies dealing with fuel economy. Nothing is being done while oil consumption and oil imports increase steadily. The nation, in effect, travels deeper and deeper into a box canyon because political leaders selfishly protect vested interests and refuse to agree on the fundamental goal.

The mistake is in the U.S. government's failure to recognize a much wider range of policy alternatives, but some states are showing the way. By creating a guaranteed market for tens of thousands of vehicles, for example, California's LEV and ZEV policies have done more to stimulate electric vehicles, and hence to reduce gasoline consumption, than any federal program since the 1970s. There are major areas where government policy can spell the difference in America's future: efforts to commercialize specific technologies, regulations that squarely address environmental threats, and tax and other fiscal policies that facilitate market-based decisions.

Green Product Development and Commercialization

The commercialization of important environmental technologies by a specific time—the year 2000, perhaps—should be announced as a clear national goal, comparable to the vow to put a man on the

Technologies and Practices Ready to Be Deployed

- Combined-cycle turbines
- Integrated gasification combined-cycle systems (IGCC)
- Circulating fluidized-bed combustion
- Pressurized fluidized-bed combustion
- Wind machines
- Solar thermal power generation
- Solar thermal water heating
- Geothermal
- Low-head hydropower
- Cogeneration, district heating, and district cooling
- Demand side technologies (low-energy windows, high-efficiency light bulbs and ballasts, high-efficiency heat pumps and furnaces, high-efficiency water heaters and refrigerators, etc.)
- Automotive improvements (aerodynamic designs, four valves per cylinder, lock-up transmissions, lightweight materials, low-resistance tires, etc.)
- Selective catalytic reduction systems
- Advanced scrubbers
- Optimized natural gas vehicles

moon. A single agency, located within or close to the White House but with links to Congress, should be placed in charge of the effort. It should initially target certain specific technologies for commercialization. Assistance should be reserved for U.S. firms, a term which does not include companies whose officers, stockholders, and headquarters are located in Japan, Germany, Sweden, or other foreign nations. One element of this program should be the development of overseas markets through a system of "green aid."

The commercialization program should encourage testing of new products to assure their suitability, quality control to assure reliability, and marketing programs to help achieve a series of successively lower cost plateaus. Low-interest or no-interest loans and loan guarantees should be available, as well as a range of tax credits aimed at rewarding the production of energy, not merely the construction of facilities. Utilities should be required to purchase

Technologies and Practices Ready to Be Commercialized

- Fuel cells for dispersed power generation
- Fuel cells for cogeneration applications
- Fuel cells for cars, buses, and locomotives
- Solar photovoltaic cells
- Chemically recuperated turbines
- Battery-powered vehicles

Technologies and Practices Requiring Development and Demonstration

- Hydrogen electrolysis
- Hydrogen transportation, storage, and use
- Advanced fuel cells

through long-term contracts a minimum percentage of their electricity from alternative energy sources, and pay the retail price for it. The elements of a commercialization program could include the establishment of efforts comparable to projects Sunshine and Moonlight, which have served the Japanese so well. These could identify nonpolluting or low-polluting sources of energy as well as identifying more efficient ways to use that energy.

Though such an effort could be located in any one of several government agencies, serious consideration should be given to assigning it to the Department of Defense. While its procurement policies sometimes result in cost overruns and other abuses, DOD deserves credit for having developed many of the technologies described in this book, despite formidable obstacles.

These programs could collaborate with private industry to support research and development, demonstrations, and commercialization by extending to prospective manufacturers tax credits, preferential financing, and other assistance designed to overcome identified barriers. An area that ought to be specially targeted is "green aid" to recruit and train employees of foreign governments and industries in the use of U.S. environmental and energy technologies, and to provide grants and loans for the purchase of such technologies.

In choosing technologies to be targeted, the highest priority should be assigned to power plants and cars. In the aggregate, these two areas account for most of the world's pollution and most of the potential profits as well. The number of technologies is relatively limited, as is the number of vendors. The top twelve carmakers account for roughly 75 percent of global production, while four companies account for virtually all of the power generation market.

Power plants, in particular, will increase in importance because electricity is inexorably displacing other fuels in a wide range of uses. In the United States, for example, electricity's share of the energy market has risen from 24.4 percent in 1970 to about 36 percent in 1989.[4] About 89.8 percent of U.S. electricity is generated from nonrenewable fuels: 54.9 percent from coal, 9.4 percent from natural gas, 3.9 percent from oil, and 21.7 percent from nuclear.[5] Given the environmental threats posed by the pollution that results from the burning of fossil fuels, the first priority must be the commercialization of technologies to produce electricity with as little pollution as possible.

Environmental Protection

An essential ingredient in the development of these technologies is the adoption of environmental protections that create a demand for them, as the recent experience in Germany illustrates. Again, the first two priorities must be cars and power plants, with specific, attainable targets in mind.[6] One defensible target would be a 50-percent reduction in emissions of carbon dioxide by the year 2010, and to eliminate them from the utility and transportation sectors altogether by the year 2050. This reduction could be achieved in a variety of ways, ranging from old-fashioned regulations to the newer "market-based" approaches currently in vogue.

Whatever the means, a phased program along the following lines would be necessary:

Effective with model year 1995, vehicles would be required to achieve corporate average carbon dioxide emission reductions from 1990 levels as follows:

1995	10 percent
2000	25 percent
2005	50 percent
2010	75 percent

Power plants and other stationary sources would be subject to comparable phased reductions, perhaps by requiring coal, oil, and natural gas generating plants with a capacity greater than 50 megawatts to cut carbon dioxide emission rates by achieving minimum levels of efficiency:

1995	35 percent
2000	50 percent
2010	75 percent

Substantial reductions in air pollution could be achieved by explicitly requiring homes and apartment buildings to be equipped with water-saving showerheads, low-energy windows, and high-density insulation as well as state-of-the-art furnaces and water heaters. Perhaps the largest energy-use and pollution reductions could arise from enactment of a take-back program comparable to Germany's or, at the least, a mandatory recycling one like Japan's.

By almost any measure, the United States consumes more energy than any other industrialized nation. A major reason for this is the design of our cities and the failure of programs that expect people to use transit systems when cars are cheaper, faster, safer, and more comfortable. There is no good reason why this should be so. Workers and students should be able to travel from home to work by public transit that is at least as fast, safe, comfortable, and cheap as cars, and this should be a national goal, with a specific maximum commuting time of, say, forty-five minutes.

Cities should also be more livable, and establishing a "green belt" around each major metropolitan area (which in the absence of some other designated boundary could be the interstate highway bypass) could help. Within that green belt, tax credit could be provided for the construction or rehabilitation of houses, offices, stores, or manufacturing facilities, and public transit systems could be improved to meet the fast-as-a-car goal.

Most would agree these are ambitious goals. The question that remains is what tools to use to advance them.

Although government regulations are now in disfavor in the United States, the contemporary American automobile owes many of its most positive qualities to these traditional "command and control" rules. Today's vehicles travel roughly twice as far on a gallon of gasoline, produce 90 percent less pollution, and save many more lives than their preregulation counterparts, all *because* of regulations. Similarly, power plants in Germany, Japan, and elsewhere operate at roughly 30 percent greater efficiency and emit about 90 percent less pollution than did those of the 1970s, again, largely due to regulation.

Although government regulations are frequently attacked as inflexible and inefficient tools (and sometimes, of necessity, they are), they do not necessarily diminish U.S. ability to compete globally, according to an industry study group, the Council on Competitiveness:

> Environmental regulations increase costs and may force weaker companies out of business but, contrary to popular perception, do not necessarily undermine U.S. competitiveness. Environmental standards are rising around the world, and often companies that are forced to meet stiffer standards gain a technological lead over companies in countries with more lax standards. Requirements to reduce waste can also result in more efficient processes that ultimately save money.[7]

Taxes and the Market

Another tool that could be employed is the tax system, which could be particularly effective in changing patterns of fuel consumption. As things now stand, oil is heavily subsidized both directly and indirectly. Eliminate those subsidies and cause the prices of coal, oil, and natural gas—and the engines that burn them—to reflect their true costs, and behavior will change. All sorts of things could be taxed: imported oil, carbon, gasoline, and large cars, for example. And all sorts of things could be untaxed or even favored with rebates: solar power, car pools, and "gas sippers," for example.

Some will object to using taxes to influence behavior on the ground that this interferes with the free market. Such objections ignore the fact that the price of energy in the United States is, and has been, maintained at unrealistically low levels by government policies that have consciously manipulated the market and consumers. At one time there were very good reasons for such policies, and they have had the desired effect: the price of gasoline, for example, is at its lowest real level in a half-century.[8] Today, however, there are equally compelling reasons for the government at the very least to step aside and allow oil prices to rise to their natural levels and to ease the pain of this transition through policies that encourage alternatives to petroleum. The truth is that the market is not, nor has it ever been, sacrosanct. Even Adam Smith, the intellectual father of the free enterprise system, reached this conclusion, supporting, for example, government-imposed monopoly under certain circumstances (such as the protection of intellectual property through patents). Initially famed for his work on social philosophy, expressed in *The Theory of Moral Sentiments*, written in 1759, rather than for theoretical economics as outlined in *The Wealth of Nations* (1776), Smith argued that government administration of a body of "positive law" was essential. "Without this precaution," explained Smith, "civil society would become a scene of bloodshed and disorder, every man revenging himself at his own hand whenever he fancied he was injured."[9]

Those who today argue that the market is somehow sacrosanct have, as often as not, been themselves willing partners and beneficiaries of earlier distortions, some of which would clearly violate the body of positive morality that Adam Smith considered essential to the achievement of a society that "flourishes and is happy."[10] For example, burning fossil fuels causes all manner of damage, but refiners, coal companies, utilities, and others pay not one whit to offset these injuries. The death toll attributable to air pollution is estimated to range upwards of 50,000 per year in the United States. In Los Angeles alone, the economic losses attributed to the adverse health effects of just two air pollutants, ozone and particulates, are estimated to be $9.4 billion a year.[11] In Japan, pollution is taxed to compensate the victims of dirty air and water. In Sweden, taxes on the sale of fuels and vehicles alike help pay not only for pollution damages, but also for the societal costs of auto accidents. In the United States, coal and oil could be taxed, while natural gas and solar, wind, and other clean forms of energy could be favored with rebates.

Other techniques that could be used are preferential rates of return (in Germany, solar, wind, and other such producers must be paid the *retail* price of the electricity which they generate), subsidized purchases (in Japan, the government provides a subsidy equal to one third of the purchase price of fuel cells), and governmentally imposed surcharges on pollution (in Sweden, emissions of sulfur dioxide, oxides of nitrogen, hydrocarbons, and carbon dioxide are taxed).

We could facilitate the development of markets by adopting nationally something comparable to California Standard Offer 4, which assured wind, solar, geothermal, cogeneration, and other technologies a ten-year market that, in turn, provided the assurance of return that investors demanded. The market itself can also be used. By pooling their resources, companies interested in reducing pollution could offer bounties for developing cleaner, more efficient products.

One of the most exciting examples of a market-based approach in the United States is the Golden Carrot program, aimed at developing a super-efficient refrigerator. Although there were federal energy-efficiency standards on the books, refrigerators much better than those required by law could be produced. The trouble was that manufacturers were unwilling to spend the money to retool to make them without some assurance that consumers would be willing to spend a few extra dollars at the time of purchase.

Meanwhile, utilities were offering bonuses to ratepayers in an attempt to induce them to buy energy-efficient appliances. A group of these utilities raised a pool of $30 million and offered it to the refrigerator manufacturer able to develop and sell units that would cut electricity consumption by at least 25 percent by boosting efficiency rather than cutting size. The winner, announced in June of 1993, was Whirlpool, which will be making the most energy-efficient and ozone-friendly mass-market refrigerator in the world—one which no doubt could compete in markets like Germany and Japan, as well as in the United States.

The government played only a leadership role in the Golden Carrot program. There were no state or federal appropriations, no new laws, no tax shifts. The same is true of a variety of other new programs at the U.S. Environmental Protection Agency, all devoted to voluntary actions.

The flagship of EPA's voluntary programs is Green Lights, which signs businesses and state and local governments to voluntary agreements to survey and upgrade their lighting systems with state-of-the-art technology that can cut lighting bills by more than half. Since lighting accounts for roughly one of every five kilowatts of electricity consumed in the United States, even small reductions in use can yield large ones in air and water pollution.

Green Lights shouldn't work, according to the classic free market theory embraced by most economists; if lighting technology really saved money, so the thinking goes, companies would already be using it. Indeed, a Green Lights contract is merely a promise that the company will do what is in its own self-interest. However, in less than two years EPA signed up more than 650 participants, including many of America's largest corporations, representing almost 3 billion square feet of office space—roughly 3 percent of the national total. Participants were expected to cut their electricity use by 12 billion kilowatt hours per year, saving $870 million.

Green Lights' success has spawned other voluntary programs. One is Energy Star, which seeks to create a market for computers that automatically "power down" when not in use, a technology pioneered for use in laptop computers. The technology adds virtually nothing to the price of a computer, but can cut its energy consumption by 80 percent. Since computers currently account for roughly 5 percent of commercial electricity consumption, this is no small accomplishment. The Energy Star program has proven so popular that it may become a de facto industry standard.

Clearly, regardless of what economic theory suggests, businesses do not necessarily recognize or act upon what's in their own best interest. Green Lights, Golden Carrot, and Energy Star all demonstrate that there are numerous subtle barriers to pollution prevention and energy efficiency improvements to be overcome even when everybody is a winner. In the case of the Golden Carrot, for example, Whirlpool wins because it receives not only a $30-million prize but also additional sales; utilities win because they can avoid constructing new power plants; and consumers win by lowering their electricity bills by 10 to 35 percent for the average household. Yet none of this would have happened without government guidance for the purpose of developing technologies to guard against environmental threats.

Sadly, in the United States even voluntary programs such as Green Lights remain the exception rather than the rule, just as companies like GEF, AT & T and SoCalGas continue to be outlyers rather than part of the mainstream. Meanwhile, America's competitors continue their quest not only for a better and safer world, but for prosperity for themselves and their children.

One of the simple facts of life is that we all must change or pay more to stay the same. For too long, the leaders of the United States have been resisting change, and the nation's people have been paying—with their jobs, their standard of living, and their future. Now it's time to change, because we can't afford not to.

A Path toward Zero Pollution

In 1990, one of the present authors commissioned a computer analysis of the pollution and cost impacts of a phased introduction of the many technologies discussed in this book. He retained an Arlington, Virginia, consulting firm, E. H. Pechan and Associates, whose past and present clients have included the U.S. Department of Energy and the U.S. Environmental Protection Agency. Pechan himself is a computer and energy expert with over twenty years of experience in the field.

The analysis relied on cost and other data provided by the government and manufacturers. It intentionally eliminated growth from the calculations because growth, whether in population or energy use, can be controlled and confuses results.

The results of that model run suggest that pollution from power plants could be reduced by between 60 and 90 percent with only modest increases in electricity costs. Major policy recommendations and projected outcomes based on those results follow:

Step 1. Adopt an energy conservation program throughout the United States, systematically replacing lights and lighting control systems, motors, and other devices that are large consumers of electricity. Based on the experience of utilities which have actually implemented such programs, electricity consumption will be reduced by roughly 10 percent. Consumers will actually save money beginning eighteen to thirty-six months after installation of the devices because of the energy saved.

Step 2. Replace all existing natural gas electric generating capacity with combined-cycle turbines operating at 50-percent efficiency. Because the combined cycle systems are more efficient, the amount of gas consumed will remain the same, but the electricity produced will increase. These increases can be used to offset reduced output at the oldest, least efficient, most highly polluting coal-fired plants.

Step 3. Replace all existing oil-fired electric generating capacity with combined-cycle turbines operating at 50 percent efficiency. Again, the new increment of electricity is used to reduce output, and hence pollution, from the older, dirty, coal-fired plants.

Step 4. Install add-on pollution control devices (scrubbers for removal of sulfur dioxide and selective catalytic reduction systems for elimination of oxides of nitrogen) on power plants more than fifteen years old, but not yet thirty. These actions will increase electricity costs, probably by an average of about 1¢ per kilowatt hour, but some of these increases will be offset as plant operators take this opportunity to make modest changes which boost overall efficiency. If the German experience translates to the United States, average power plant efficiency will rise from about 34 percent to about 38 percent. This gain, combined with the use of pollution control devices, will lower emissions of all major pollutants.

Step 5. Convert all coal-fired power plants over thirty years old to one of the coal gasification technologies such as Cool Water (IGCC). Emissions of most pollutants other than carbon dioxide will drop 90 to 98 percent. Because the new plants will be operating at efficiencies of up to 44 percent versus the current average of 34 percent, emissions of carbon dioxide will drop by roughly 30 percent. Operating costs should remain roughly level or perhaps drop due to reduced coal consumption.

Step 6. Target one half of the total capacity for conversion to cogeneration. By channeling waste heat to nearby industries, or using it for heating and cooling of offices and apartments, overall efficiency should be raised to at least 75 percent and, in some cases, reach the 90 percent routinely achieved in Europe.

Step 7. Starting between the years 2000 and 2005, begin replacing conventional generating units with fuel cells or advanced aircraft-derivative turbines. These can operate at efficiencies of roughly 60 percent, and in cogeneration modes they allow virtually all of a fuel's energy to be used.

Step 8. As zero-polluting generating technologies become progressively more competitive, phase them in. Wind power, which can already generate electricity at prices competitive with coal, should be deployed beginning in the early 1990s. Solar thermal power stations should become competitive by the mid-1990s. Solar photovoltaics, which are already competitive for remote installations, should begin to compete for baseload electricity generation beginning around the year 2000, perhaps earlier for some uses, especially if coal and oil prices start to reflect the true costs of using these fuels. Nuclear stations using inherently safe reactor technology can begin to go on line at the same time.

Step 9. Convert the entire utility system to zero-polluting sources of electricity, which can be used to electrolyze water for hydrogen to fuel cars, planes, trains, and trucks.

Would such a program be easy? No, but it might not be all that difficult either. Certainly the United States has confronted greater challenges, ranging from electrification of the nation to the construction of the intercontinental rail and highway systems. The issue then, as now, is not whether undertaking and completing such formidable tasks will be easy, but whether they are worthwhile: Will the nation be better or worse for having done them?

Perhaps the nation will be better off by continuing to waste two of every three pounds of coal in its power plants and four of every five gallons of gasoline in its cars—but it seems unlikely. And perhaps the nation will prosper for having completely abandoned solar electricity, fuel cells, and super-efficient turbines to the Japanese and Germans, though it's difficult to see how. Then again, Americans may prefer to spend increasingly large shares of their money and time on clogged freeways, fighting their way from one fringe of a smog-choked city to another—but it seems doubtful. Indeed, the likely—the virtually certain—preference of most citizens would be to breathe cleaner air, consume less energy, and waste less time and money. That, in our judgment, is what we have proposed.

Some will resist the very notion of such an undertaking on the grounds that some will suffer because of it. They confuse change with sacrifice. Certainly it was a change, for example, to eradicate polio—and for the manufacturers of iron lungs it was no doubt a sacrifice as well—but for the nation as a whole, especially its children, it was a great victory.

Others will quarrel with the details of what we have outlined. Fine. There are many options, and what's important is for the nation to agree upon goals, and do so quickly, rather than waste more time in a bitter and divisive debate over the means. It is with that hope, and in that spirit, that we have written this book. There is a better world than the one we live in today, and for the sake of our nation and its children, we hope that America will be the nation that invents that better world.

Epilogue

> Since the Cold War . . . we've been looking for some new enemy. Well, maybe the new advesary is the oldest one of all. And we think of trillions that were spent on exquisite systems of alert and defense against some enemy; maybe it's time to think of turning resouces to alert and defense against the forces of nature.
>
> —DANIEL SCHORR, National Public Radio

After the elections of 1992, and under the leadership of a new president, the United States once again began to embrace the prospect of new environmental regulations and new technologies for meeting them. Looking to the future instead of clinging to the past, the new administration promised to challenge Japan and Germany in the race for supremacy in a new generation of clean, efficient, and safe technologies.

On Earth Day, President Bill Clinton announced that the United States would cap emissions of air pollutants that cause global warming. Although Elizabeth Barratt-Brown of the Natural Resources Defense Council cautioned that "the proof is always in the pudding," she nonetheless lauded Clinton: "Today he sent very clear marching orders to the U.S. government on these issues. They're very important issues in and of themselves but they're also enormously symbolic issues that signal a new era of leadership on global environmental challenges."[1]

Later that spring, President Clinton announced that he would seek to curb the ballooning federal deficit by relying on a new energy tax, one that would not only raise revenue but also encourage conservation and the development of cleaner fuels. In the fall, he joined with the chief executives of the Big Three automakers to announce a new Detroit-Washington partnership to develop what was quickly dubbed by the *New York Times* the "government dream car."[2] Compared to today's models, it would travel three times as far on a gallon of gas, produce only a fraction of the air pollution, and be largely recyclable.

Finally, President Clinton unveiled the Climate Change Action Plan at a Rose Garden ceremony on October 19, 1993, a cloudy fall day exactly twenty years after the start of the Arab oil embargo. The plan, said Clinton, was designed to return U.S. emissions of Greenhouse gases to 1990 levels and thus fulfill his campaign promise to make the fight against global warming a domestic policy priority—a pledge that had helped him garner support from virtually all of the nation's major environmental groups.

On the surface, these and other actions seem to closely follow the prescriptions provided in the previous chapter for altering the course of the country's future. And without doubt there has been plenty of change. The rhetoric of the new president and his appointees *is* different; instead of denying the existence of global warming, for example, they have joined the other industrial nations in saying that prudence demands actions now, without waiting for irreversible damage. The people are different, too; Clinton's appointees are honestly worrying about the challenges confronting America and attempting to develop solutions. And there's the problem—the solutions.

The president, together with the leaders of virtually all the world's nations, has described challenges equal to any that have ever confronted the United States. In defining solutions, however, he and his appointees—perhaps because they're daunted by the harsh political realities of a Washington power structure still dominated by those who oppose such change—seem gripped by timidity. Some say they're merely acknowledging the realities of Washington; others say that the problems are no longer being trivialized but the solutions are. There's truth in both positions.

Consider, for example, the "dream car" initiative. Bear in mind that the United States is more dependent on Persian Gulf oil today than it was at the time of the original oil embargo in 1973 and that the nation spends more money on imported oil than it does on imported cars. Cars are the chief cause of the urban smog that so chokes U.S. cities that roughly one of every two Americans is forced to breathe air that's at least unhealthy and sometimes lethal. One of every twenty pounds of carbon dioxide in the world comes from a U.S. tailpipe—more than *all* of Japan or Germany's carbon dioxide emissions.

Yet the goal of the dream car initiative is to mobilize all of the engineering talent of the United States to produce only a single version of a car sometime in the next century. It is a voluntary program, with no new funding, no specific deadlines, and no clearly defined objectives. President Clinton, in announcing the project, said, "We intend to do nothing less than redefine the world car of the next century, to propel the auto industry to the forefront of world automobile production."[3] But how?

In Japan, pronouncements such as this one would have been tempered. They also would be undergirded by detailed plans containing explicit targets and deadlines, and accompanied by multi-year budgets and specific government actions designed to implement the strategy. In Germany, the elaborate planning of Japan might have been absent, but the regulatory requirements would have been sharp and well defined. All of these were noticeably missing from the dream car initiative. Pressed to describe what technologies the car might actually employ, for example, a Detroit official couldn't. His feeble explanation: "There is no silver bullet."[4] Similarly, a White House press release proclaimed the initiative to be "a historic new partnership" but described the deadline for results vaguely as "within approximately a decade."[5] The agreement, conceded the *New York Times*, was "only a goal, and not a clear one at that."[6]

In Japan, such an effort would be closely managed and overseen, with regular meetings. In Germany, the auto companies would have been left to their own devices, but a failure to meet the deadlines and other requirements would not be countenanced and the companies would know it. Not so with the dream car project. "Maybe every year or two we should have a kind of a review by peers of technical progress," said Clinton's science advisor, John Gibbons.[7] Perhaps most importantly, there were no concrete commitments whatsoever from the chief executives of General Motors, Ford, or Chrysler.

Such arrangements are not lightly entered into in Japan, partly because failing to achieve announced objectives is cause for shame and disgrace, for both the government and the companies. Yet in the case of the U.S. initiative, evidence began to mount that instead of being a sincere effort to develop radically new vehicles, it was seen, by Detroit at least, as a means of fending off regulation. Exactly three weeks after the initiative was announced, the front page of the *Washington Post* revealed that Detroit's Big Three carmakers were pressuring Clinton administration officials to bar the adoption of California's tailpipe standards by states in the Northeast. The auto industry was, said the *Post*, "using its new alliance with the White House" in efforts to overturn the proposed standards, which had already been adopted by New York, Massachusetts, Maine, Maryland, and several other states. "Before [the dream car initiative] we didn't have much credibility in Washington," said Ford chairman Harold Poling. "We now have the government looking at the same data we are looking at; they are looking at our research."[8] (Evidence of the success of the Big Three's strategy was found in the January 25, 1994, *Wall Street Journal*, which said that "the biggest payoff from the supercar effort may be a relaxation of a broad array of costly regulations." Dr. Mary Good, head of Clinton's dream car initiative, commented that the *Journal* "equates the search for zero-emission vehicles with the quest for 'perpetual motion machines.' That's a view certain to cheer Detroit.")

Of course, it would be understandable if the dream car initiative fell somewhat short of the mark. Detroit's carmakers are notoriously slippery and they're defended by some of Washington's most powerful politicians, including John Dingell, the Michigan Democrat who chairs the powerful House Committee on Energy and Commerce. An unflinching ally of the auto industry and foe of anything that smacks of environmental protection, Dingell is a take-no-prisoners adversary who once said that when hunting he always uses "the biggest rifle or the biggest shotgun" he can take, adding,

"When I shoot something I want it to go down. I don't want any cripples walking around."[9] Clinton would be understandably reluctant to upset Dingell, who would have much to say about the health care reform package that had become the president's first domestic priority.

Still, this gap between the size of the challenge and the magnitude of the solution can be seen in other proposals as well. An example is the Climate Change Action Plan. Described as a program to return U.S. carbon dioxide emissions to their 1990 levels by the turn of the century, the plan consists of fifty measures ranging from tree planting to increased use of hydropower, almost all of them voluntary. "Painfully modest" was the description offered by the *National Journal*.[10] Though Clinton supported tightening federal fuel-economy standards for cars during the presidential campaign, the Climate Action Plan rejected this means of reducing pollution. Neither is there a program for aggressively deploying any of the new power plant technologies we have described. Similarly omitted are measures to coax alternative energy systems onto the market, other than a limited expansion of hydropower.[11]

Not all of this can justly be laid at the doorstep of the Clinton administration. To its credit, this administration's first major proposal was to reduce the federal deficit by adopting a new tax on energy. Such a measure would almost certainly have stimulated the market for new technologies, yet opposition quickly mounted in Congress, and, faced with resistance from industries ranging from international oil companies to recreational vehicle manufacturers, Clinton backed away. In doing so, he was essentially recognizing the almost inevitable outcome, given the dependence of Congress on special interest money and the unwillingness of those narrow interests to support measures for the common good. The result was a gasoline tax increase of 4.3¢ per gallon, too little to influence the behavior of all but a few motorists and vehicle buyers. In short, while both the Congress and the administration are now willing to admit the existence of the imperatives this book discusses, neither is apparently able to adopt solutions on the requisite scale.

Of course the new rhetoric is important in and of itself. The words and statements of leaders can send invaluable signals to industry and citizens alike. Moreover, such rhetoric can pave the way for later deeds. But actual accomplishments are most persuasive, and sadly, the deeds of U.S. policymakers are falling short of their words—and this disjuncture couldn't come at a worse time.

To the extent that the United States has retained its ability to compete in the global market for cleaner cars, power plants, and other products, it has been because of the resolute policies of California. But with the conclusion of the Cold War and sharp reductions in defense spending, industries in California have used the state's economic downturn as a weapon against its energy and environmental policies. "Businesses are clamoring for less regulation, not more, because they contend that over-regulation combined with a bad economy are hurting their ability to operate," said the *Los Angeles Times*.[12]

In Los Angeles, industry-financed campaigns have managed to force three of the most consistent and articulate advocates of environmental technologies off the board of the South Coast Air Quality Management District. Henry Morgan, a moderate Republican who chaired the board's Technology Committee, was defeated in a recall election heavily influenced by funds from industries unhappy with his support of air pollution rules.[13] Constant hammering by the same groups forced Larry Berg, the board's most outspoken and articulate member, to resign because he was, after a decade, simply too tired to continue the fight. Sabrina Shiller's appointment expired after industry pressure was brought to bear. And Hank Wedaa, the middle-of-the-road Republican who has chaired the board since 1991 and who has championed the commercialization of fuel cells, is the target of an industry-backed campaign so powerful that it succeeded in passing a new state law specifically aimed at ousting him from the board. A "marked man," said the *Los Angeles Times* of Wedaa, reporting that law's passage was "a victory for businesses eager to loosen strict air pollution rules."[14]

In Sacramento, the news has been equally bleak. Pete Wilson ran successfully for governor as a Republican moderate who supported stringent environmental controls. But he quickly moved to rein in the state's independent environmental agencies, especially the California Air Resources Board, through the creation of a new, overarching California Environmental Protection Agency (Cal/EPA). Jim Strock, a former appointee of both Bush and Reagan chosen by Wilson to head Cal/EPA, gradually wrested control of CARB away from its longtime chair, Jan Sharpless, who was then removed from office by Wilson. "[C]ARB's regulatory powers have long been challenged by manufacturers and others who wield influence with the Republican governor," reported the Associated Press in an article on Sharpless's departure.[15] Though Sharpless was removed from CARB, she was immediately appointed by Wilson to the California Energy Commission (CEC), the innovative agency that has helped launch the alternative fuel revolution in California. Then Wilson proposed to abolish the commission.

David Freeman, the populist head of the Sacramento Municipal Utility District, who was credited by the *Sacramento Bee* with "bringing SMUD back from the brink of collapse," retired on February 1, 1994, to become president and chief executive of the New York Power Authority.[16] Change was also afoot at Pacific Gas and Electric, the San Francisco utility that pioneered solar, wind, and geothermal power while establishing what many consider to be the nation's most aggressive and focused program for "demand side management," or conservation. Faced with a refusal by Governor Wilson's appointees to the California Public Utility Commision to allow the costs of developing solar, wind, and other new forms of generation to be recouped through rates, the company has decided to buy its electricity from others rather than build new power plants.[17] This eliminates what some believe has been the single most potent and effective force for utility innovation in the United States.

Matters were no better in the California legislature, where the long series of laws that had helped propel the state to the forefront of environmental technology had been enacted. The Assembly Natural Resources Committee approved proposals to "streamline" the California Environmental Quality Act (CEQA), slash the fees dedicated to supporting technology development efforts at the South Coast Air Quality Management District, and roll back deadlines for increased use of recycled goods. The reason: "constant pounding by business lobbyists," in the words of the *Orange County Business Journal*.[18]

These changes, especially the departure of Sharpless, who was an outspoken proponent of the California zero-emission-vehicle requirements, opens the door to a relaxation of tailpipe rules. Buoyed by their successes in Washington, D.C., and the Northeast, U.S. automakers mounted "sharp new attacks on California's mandate," according to the *Los Angeles Times*.[19] Ford has appealed directly to Wilson, although it's General Motors that is leading what one of the carmakers' own internal memoranda described as "a broad, consistent and joint plan" for overturning the ZEV requirements, a plan whose targets include not only the governor, but also Vice President Gore.[20]

For nearly a decade and a half, California's resolve and foresight cushioned the blows that were dealt by the administrations of presidents Bush and Reagan as well as a compliant Congress. But with the dramatic changes in Los Angeles and Sacramento, California's leadership can no longer be taken for granted. And indeed it shouldn't be, because the responsibility rightly belongs in Washington.

We said in the previous chapter, and it is worth restating, that today U.S. environmental regulations are too lax, energy prices too low, investment capital too expensive, and governmental assistance too limited to stimulate the commercial innovation that characterized the 1960s and 1970s. The first step toward restoring U.S. competitiveness is to admit these harsh and unpleasant realities, then commit ourselves to solutions.

That brings this drama back full circle. The government that got us into this mess must now extricate us. It is unrealistic to expect that problems that required years to create will be solved in months. Still, the first few steps have been at best shaky and halting. That won't do, for there is no room for half-steps or mistakes, because Japan and Germany will make few if any. Temporarily, both of those nations are preoccupied by other matters—Germany by the massive task of reunifying the former East and West, Japan by its worsening economy and the faltering shift of power away from the World War II generation. Still, the programs of these nations remain on track, with their vision and vigor largely undiminished.

Indeed, this may be the most disheartening aspect of our nation's recent slow but steady retreat from a position of leadership in environmental technologies and policies. For two centuries, the United States has rightly prided itself as a nation of vision and courage, sustained by a faith in the resourcefulness and resiliency of its people. Today, in contrast, our elected officials reject the notion that American ingenuity will find solutions when forced by necessity to do so. They in effect refuse to acknowledge problems because of their own inability to imagine the solutions. To assume that people who developed airplanes invisible to radar, submarines silent to sonar, and bombs capable of dropping through a doorway from a height of ten miles cannot build practical electric cars and clean-burning power plants condemns us to forever avoid opportunity rather than confront challenges.

America's leaders, especially those in the Congress and in business, still fail to understand that what is at stake is the survival of the United States as an industrial power in the twenty-first century. Rhetoric alone will not put the nation back on track. What is required is either clear, explicit guidance in the form of rules, regulations, taxes, or other specific and unambiguous policies, or focused, fully funded and coordinated technology development programs. The United States has successfully employed both approaches in the past and could do so again, just as our industrial competitors are doing now. In either event, the first step is simple, old-fashioned leadership that produces actions, not just words.

Notes

Introduction

1. AT & T, "A Healthy Balance: AT & T Environment and Safety Report 1992" (Basking Ridge, N.J.: AT & T, 1993), 6–7.

2. ECD's production line produces 14-inch-wide, 2,500-foot-long rolls of amorphous silicon alloy solar cells deposited uniformly on a 5-mil. stainless steel substrate. The production yield, uniformity, and consistency of the process have been demonstrated to be high and reproducible. The efficiency of the solar cells produced will be roughly 13 percent. ECD officials say the continuous roll-to-roll process can be readily scaled up to a 100 MW production line through the use of proven, straightforward engineering improvements. The production costs of photovoltaic (PV) modules produced in a 100 MW continuous roll-to-roll process has been calculated to be less than a dollar per peak watt. The cost of electricity generated in PV power plants using amorphous silicon alloy solar cells produced in such a plant will therefore become competitive with the cost of electricity produced in conventional utility power plants that utilize coal, gas, or oil.

3. Canon and ECD had dealt with one another before. Prior to the photovoltaic joint venture, ECD had developed a revolutionary drum for Canon photocopiers, using photoreceptor technology.

4. See Thomas H. Lee and Proctor P. Reid, eds., *National Interests in an Age of Global Technology* (Washington, D.C.: National Academy Press, 1991), 124.

5. Ibid., 123. However, even the advantage in clean coal technology will likely vanish, since the acid rain control legislation passed in 1990 consciously encourages fuel switching as the means of compliance, rather than installation of control technology.

6. Ibid., 124–25.

Chapter One. Germany's Miracle

1. Much of the information contained in this chapter is based on numerous interviews and site visits conducted during visits to Germany from 1986 through 1993. Among the most informative of these was a seminar and study tour June 3–12, 1991, sponsored by the United Nations' Economic Commission for Europe and organized by the German environmental protection agency, the Umweltbundesamt (UBA). Some of the materials were collected while on assignment for *International Wildlife* magazine. The authors are particularly grateful to Dr. Axel Friedrich, Dr. Norbert Haug, and Dr. Bernd Schärer of the UBA for their expert advice and assistance during these visits and the preparation of this book, and to the editor of *International Wildlife*, Jon Fisher, for his support, encouragement, and interest.

2. Council on Competitiveness, *German Technology Policy: Incentive for Industrial Innovation* (Washington, D.C.: 1992).

3. Umweltbundesamt, "Standortfaktor Umweltschutz," 28 February 1993, Berlin.

4. Christopher Doiron, New Brunswick Power, interview, 27 August 1993.

5. Roger Gale, interview, 20 March 1992.

6. Conrad von Moltke, interview, 20 January 1992.

7. Because emissions of sulfur dioxide and oxides of nitrogen are largely uncontrolled in the former East German states, they are among the highest in Europe on either a per capita or per GNP basis. SO_2 emissions now total about 5.4 million tons—thirteen times the level in the former West Germany. However, during the next two years the formerly East German power plants are faced with the same choice that confronted their Western counterparts in 1983: by July 1992 each plant must commit to shutting down or curbing its air pollution in accordance with the ordinance. Plants can opt to shut down rather than comply with the law, but in such a case the facility is barred from continuing to operate for more than 3,000 hours or after July 1, 1996. Plants which opt to continue in operation, and therefore to install pollution control equipment, will buy German systems.

8. Wolfgang Hilger, quoted in Curtis A. Moore, "Down Germany's Road to a Clean Tomorrow," *International Wildlife*, September/October 1992, 24–28.

9. Germany's continued commitment to improved environmental technology despite poor economic times is illustrated by two items that ap-

peared in the *Washington Post* of January 20, 1994. An article on Europe's economic problems noted that German unemployment in December 1993 was at 9 percent, up more than 1 percent from the year before, and its expected economic growth in 1994 is less than one percent (Hobart Rowen, "Europe in the Mire of Joblessness," *Washington Post*, 20 January 1994, A20). Yet an advertisement for Mercedes Benz in the same paper boasted that its new E-300 model reduces hydrocarbon emissions by 43 percent and carbon monoxide by as much as 93 percent, making it "the only diesel automobile to successfully pass the emission standards of all 50 states" (ibid., D12).

10. See Gregory S. Wetstone and Armin Rosencranz, *Acid Rain in Europe and North America: National Response to an International Problem* (Washington, D.C.: Environmental Law Institute, 1983), 84. The German standard was 140 micrograms per cubic meter (ug/m3), compared to the European Community (EC) standard of 80–120 ug/m3 and the World Health Organization's recommendation of 40–60 ug/m3.

11. E. Rehbinder, "Implementation of Air Pollution Control Programs Under the Law of the Federal Republic of Germany," *Air Pollution Control: National and International Perspectives* (Washington, D.C.: American Bar Association, 1980), 31.

12. Peter Schütt and Ellis B. Cowling, "Waldsterben a General Decline of Forests in Central Europe: Symptoms, Development and Possible Causes," *Plant Disease* 69, no. 7 (July 1985): 548. *Waldsterben* wasn't confined to Germany. Switzerland, Austria, Sweden, and other nations also found widespread forest death, which led to accelerated antipollution programs in those nations as well.

13. For "hard-coal" plants, the standard for oxides of nitrogen in Germany has effectively required the installation of SCR systems. However, plants fired with lignite have been able to meet the standard without the use of SCR through the adoption of a variety of "primary" measures such as burner modifications and exhaust gas recirculation. By 1991, about 150,000 megawatts of capacity had been equipped with primary measures, and another 90,000 megawatts with SCR.

14. The most frequently adopted system is so-called wet scrubbing, which is used in systems accounting for roughly 42,000 megawatts of electric capacity in Europe. These systems account for 100 percent of the flue gas desulphurization (FGD) capacity in the Netherlands, and about 89 percent in Germany, and they routinely operate at removal efficiencies of higher than 90 percent. In Turkey, an FGD system is operating at 98-percent removal because of the high sulfur content of the lignite being

burned there (about 7 percent by weight). Scrubbing systems utilizing lime or limestone are the most common, accounting for 87 percent of Germany's installed capacity. Systems are, or soon will be, installed in the United Kingdom, Italy, and Spain as well.

15. Helmut Weidner, *Air Pollution Control Strategies and Policies in the Federal Republic of Germany: Laws, Regulations, Implementation and Principal Shortcomings*, 95 Edition Sigma Bohn (Berlin, 1986), 90.

16. The action was ridiculed by some, who claimed without evidence that Germans were merely pocketing the unlevied taxes, while throwing the converters in the cars' trunks. This was a favorite criticism of William Chapman, a veteran Washington lobbyist for General Motors, when seeking to deflect momentum for a tightening of U.S. tailpipe standards.

17. See Michael P. Walsh, "Europe on the Verge of State of the Art Auto Standards," *Car Lines* 1, no. 2 (April 1989): 2; and "European Community Environment Ministers Agree on New Emission Levels for Small Cars," *International Environment Reporter* 12, no. 6 (14 June 1989): 283.

18. Michael P. Walsh, international motor vehicle emissions consultant, Washington, D.C., personal communication, January 1990.

19. Based on these standards, about 54,000 megawatts of electric capacity had to be retrofitted with controls for the reduction of sulfur dioxide and oxides of nitrogen. The vast majority of plants opted in favor of wet scrubbers for the removal of sulfur dioxide.

There are about 170 power plants in Germany equipped with operational scrubbers in compliance with the 1983 federal law. About 140 of these plants are also equipped with selective catalytic reduction. However, neither scrubbers nor SCR are required per se by the law, which establishes numeric standards that can be met by the source as it wishes. The utility sector has, pursuant to the Large Firing Ordinance of 1983, reduced its emissions of SO_2 by more than 90 percent, at a cost of about DM 40 billion, and its emissions of NO_x by 80 to 85 percent.

20. See the report of the Enquete Commission of the German Bundestag, *Protecting the Earth's Atmosphere*, translated into English by Wolfgang Fehlberg and Monica Ulla-Fehlberg (published by the Bundestag, Bonn, 1991), 164–69. Because transportation issues cut across almost every other aspect of society (work patterns and the location of industrial facilities, for example) German planning is focusing on what the government calls fundamental reorientation. The objective is to develop a regional planning and transport system which should be as environmentally and climatically sound as possible. Specifically, the government's objectives include measures to "avoid, re-route, guide, and reduce traffic," and programs to develop better technology.

21. Evert Anderson, "Trains of Tomorrow," presentation at a conference entitled "Global Survival: Sustainable Development in the Networking World," Swedish Royal Institute of Technology, Stockholm, 9–10 August 1993. This calculation assumes a load factor of 40 percent on a train hauled by an ordinary electric locomotive, compared to a five-seat car with a load factor of 40 percent traveling at about 60 miles per hour, or a jet airplane with a load factor of 60 percent.

22. Ibid.

23. Clifford Black, director of media relations, Amtrak, personal communication, 11 January 1994.

24. Anderson, "Trains of Tomorrow."

25. Much of the responsibility for implementing Germany's aggressive environmental program is lodged with the Lander, or states. Energy agencies have been established in several federal states (e.g., North-Rhine Westphalia, Saarland, Lower Saxony, Hesse) to facilitate energy efficiency, in particular in the sector of small-scale users (mainly small and medium-sized enterprises, as well as public buildings), by offering advice and third-party funds and by encouraging energy conservation.

In ten federal states funds have been earmarked specifically for the promotion of cogeneration (of heat and electricity from the same fuel). In their energy management concepts, the regional and local authorities involved have outlined energy conservation possibilities for their areas of supply, and they have specified measures designed to make use of the energy conservation potentials identified. Some local authorities have defined emission reduction targets for greenhouse gases in the areas for which they are responsible. The city of Schwerte, for instance, has decided to reduce CO_2 emissions by 30 percent (relative to 1989) by the year 2000.

26. Germany has also launched the world's most ambitious program to capture and recycle the CFCs contained in refrigerators. When an old refrigerator dies, it is hauled at no charge to one of eleven centers where its compressor's CFCs are drained. Then the sheet metal is carefully peeled away, exposing the CFC-laden styrofoam, which is gingerly extracted and crushed beneath exhaust hoods that capture the ozone-destroying gases. The gases are then piped to barrels for storage and recycling.

27. See Enquete Commission, *Protecting the Earth's Atmosphere*, 164–169. Germany decided on June 13, 1990—well before most other industrialized nations—to adopt the target of reducing its CO_2 emissions by 25 percent from 1987 levels. A few months later, after unification of the two Germanies, the government had not confirmed the 25-percent reduction target for the old federal states, but did impose an even higher target of 30 percent for the former East. The government's decisions were based on

recommendations made by a highly influential committee of the German Bundestag, or parliament, the Enquete Commission on preventive measures to protect the earth's atmosphere.

By 1992, the following measures had been taken or were planned at the federal level: (1) A new Bundestarifordnung Elektrizität (federal survey of electricity charges) was adopted to eliminate the practice of discounting electricity rates for large-volume consumers like industries. The new system, which entered into force in early 1990, requires rates which are more linearized. (2) The Stromeinspeisungsgesetz (electricity sales act) requires public utilities to purchase electricity generated from renewable energy sources, and establishes minimum compensation to be paid for such electricity. (3) The Energiewirtschaftsgesetz (energy management act) will be amended to include sustainable management of resources and environmental protection in the act's catalogue of objectives. (4) An amendment to the Wärmeschutzverordnung (ordinance on heat conservation) is being drafted to develop "low-energy house standards" for all new buildings.

28. "A Wall of Waste," *Economist*, 30 November 1991, 73.

29. Whenever it believes market forces will do a better or cheaper job than traditional "command and control" mechanisms (in which the government issues detailed orders and regulations), the German government has tried to use them; like most German environmental laws, the take-back program encourages industries to develop their own solutions on the theory that private enterprise can often work better and more cheaply than government. In the case of the take-back program, the law's effect has been to create two somewhat parallel systems for dealing with waste—one operated by government, the second run by the industries themselves—the Duales System Deutschland.

30. Cynthia Pollock Shea, "Package Recycling Laws," *BioCycle*, June 1992, 56–58.

31. "Ab Sofort. Alle Grundig Fernseher Mit Recycling Garantie," *Frankfurter Allgemeine*, 10 October 1992.

32. Bundesverban der Deutschen Industrie e.V., *International Environmental Policy: Perspectives 2000* (Cologne: 1992), 37.

33. Dean E. Murphy, "Germany's Recycling Nightmare," *Los Angeles Times*, 12 September 1993, D5.

34. Ibid.

35. Umweltbundesamt, *The Environmental Label Introduces Itself* (Berlin: 1991), 5–6. Labels are awarded only after a cradle-to-grave scrutiny of the product is overseen by the government and accompanied by lengthy testing and hearings. The findings are then reviewed by an independent jury of eleven voting members, including representatives of the manufacturers,

consumers, and environmental groups. To receive the right to display the Blue Angel, manufacturers must pay a yearly fee of between $21,000 and $250,000, depending on the product's annual sales. These fees support the Angel program.

36. Friedhelm Mensing, "Environmental Conservation: A Question of Man's Survival," Inter Nations Press special report SO 1–1990 (Bonn). Much of Germany's green aid is not directly related to the marketing of German goods. Liberians, for example, are being taught how to mill, plane, and dry their own lumber so that more income can be generated by felling fewer trees. In Pakistan, India, and Afghanistan, 21,000 improved versions of the traditional tandoors, or wood stoves, have been distributed. By nearly quadrupling the number of flatbreads which can be baked per unit of wood, the stoves can slash tree-cutting for fuel by roughly 75 percent.

37. "European Report," *Wall Street Journal Europe*, 11 August 1993, 1.

38. Michael Porter, "America's Green Strategy," *Scientific American*, April 1991, 168.

Chapter Two. Japanese Opportunism

1. Much of the information relating to Chiba was gathered by author Curtis Moore during a visit to the works in December 1985. Details regarding the production processes are drawn from an undated and untitled video provided by Kawasaki Steel at that time (produced by Iwanami Production, Inc).

2. For general information, see Keidanren, "Japan's Industries Work for Conservation of Global Environment" (1992); and Industrial Pollution Control Association of Japan, *Environmental Protection in the Industrial Sector in Japan* (1983). From 1970 to 1980, Japanese steel plants reduced SO_x emissions by 60 percent, NO_x emissions by 30 percent, and dust by more than 80 percent. In the peak year of investment in pollution control in 1976, the iron and steel industry spent over 20 percent of its total investment in plants and equipment.

3. "If We Knew Everything Then We Know Now," *Business Week*, 15 January 1990, 54.

4. Takefumi Fukumizu, "Our Threatened Environment: Challenges and Opportunities," presentation at the Conference on Clean Air Business Opportunities, sponsored by the South Coast Air Quality Management District, Newport Beach, California, 1 October 1992 (meeting hereafter referred to as the Newport Beach Conference).

5. Ibid.

6. Ibid.

7. Ibid., presenting data provided by the Organization for Economic Cooperation and Development (OECD).

8. Ibid.

9. Study Group for Global Environment and Economics, Japan Environment Agency, "Pollution in Japan—Our Tragic Experiences" (July 1991), 20.

10. Masatoshi Furuichi, "Environmental Issues in the Global Community, Striving for Clean Air and Power: The Japanese Experience," presentation to the Environmental Technology Seminar, Long Beach, California, 30 November 1989. It has been estimated that Japan is responsible for about 5 percent of the world's carbon dioxide emissions but about 14 percent of the world's gross production.

11. Michael Porter, "America's Green Strategy," *Scientific American*, April 1991, 168.

12. Frederick S. Myers, "Japan Bids for Global Leadership in Clean Industry", *Science* 256 (22 May 1992): 1144–45.

13. Quoted ibid.

14. See William Cline, *The Economics of Global Warming* (Washington, D.C.: International Institute for Economics, 1992).

15. Quoted in Myers, "Japan Bids," 1144–45.

16. Hideo Suzuki, "The IEA Kyoto Conference Keynote Speech Outline," presented at the conference on Technology Responses to Global Environmental Challenges, Kyoto, Japan, 6–8 November 1991.

17. Memorandum from Richard L. Lawson, president of the National Coal Association, 15 May 1991. The Information Council on the Environment was a public relations advertising campaign to test-market newspaper and radio ads in Bowling Green, Kentucky; Fargo, North Dakota; and Flagstaff, Arizona, during May 1991. The campaign was funded largely by members of the National Coal Association, fifteen of whose members contributed a total of at least $75,000 to the project. Despite the name, selected from many others by a public relations agency, there was no council.

18. Federation of Electric Power Companies, "The Electric Power Industry and Environmental Issues" (Tokyo: Keidanren Kaikan, undated).

19. Ibid., 5.

20. Ibid., 5–6. Other measures listed included the following: active introduction of highly efficient liquid natural gas complex power production; increased development of technology for molten carbonate and solid oxide fuel cells; development of power production technology through pressurized flow bed complexes and technology incorporating a high de-

gree of efficiency; steady introduction of hydroelectric and geothermal power; introduction of a policy on solar and wind-generated energy; by the middle of the twenty-first century, cooperation at the national level to develop technology for nuclear fusion and hydrogen-based energy; development and popularization of "as yet [un]used power sources" such as exhaust gas from heat pumps, power generated by the flow of river water, and heat generated by the subway system; development and popularization of cogeneration systems; and development and popularization of energy-efficient machinery and systems to be used with electric automobiles and "super-heat pumps."

21. Ibid.

22. See Alan Miller and Curtis Moore, "Japan and the Global Environment" (College Park, Md.: Center for Global Change, 1991).

23. See "Japan's Foreign Aid Program: Setting Priorities, Policies in 1992," JEI report no. 46A, 11 December 1992. (JEI stands for Japan Economic Institute of America.)

24. Choy, "Japan's Energy Policy: A Powerful Example?" JEI Report no. 23A, 25 June 1993.

25. Japan's achievements in the field of environmental technology are a not coincidental component of its larger record of accomplishments in economic growth through a national commitment to the commercial application of advanced technology. As the economist C. Fred Bergsten notes, one of the most important reasons for Japan's transformation from postwar basket case to economic superpower "has been the growing prowess of Japanese firms in high technology industries. As little as two decades ago, Japanese firms were generally considered to be marginal players in these industries. Today, by contrast, Japanese firms are now the acknowledged technological leaders in many advanced sectors" (C. Fred Bergsten, preface to Thomas Arrison, C. Fred Bergsten, Edward Graham, and Martha Harris, eds., *Japan's Growing Technological Capability: Implications for the U.S. Economy* [Washington, D.C.: National Academy Press, 1992], v).

26. Louise D. Jacobs and Leigh Harris, *Public-Private Partnerships in Environmental Protection: Use of Public-Private Partnerships in the U.S. and Japan to Avert Global Warming* (Washington, D.C.: Council of State Governments, 1992). The New Earth 21 Action Program is described as "an international action program with a long term point of view which is based on two main ideas: the development and domestic disseminations . . . of innovative technology, in the fields of energy and environment, and the wide-scale transfer of technology in these fields to the world."

27. Myers, "Japan Bids," 1144–45.

28. Unless otherwise indicated, the information contained in this section is drawn from a Japanese government brochure, *NEDO: New Energy and Industrial Technology Development Organization,* which is distributed by the government of Japan and was printed in March 1991 (hereafter referred to as *NEDO*). Although the body was officially renamed the New Energy Development and Industrial Technology Development Organization in 1988, it is still almost invariably referred to as NEDO.

29. Kunisukwe Konno, NEDO director general, Alcohol and Biomass Energy Department, and director, Overseas Affairs Division, Policy Planning Department, personal communication, 12 November 1991. Kunisukwe Konno estimates that there are about a hundred other quasi-governmental organizations.

30. A large number of NEDO personnel are assigned to assist with retraining and other services for coal workers displaced by a shutdown of the industry—another possible lesson for the United States.

31. Kunisukwe Konno, personal communication, 12 November 1991.

32. Ibid.

33. Kunio Maeda, NEDO deputy director, Overseas Affairs Division, Policy Planning Department, personal communication, 6 November 1991.

34. R. Anahara, chief engineer, Fuji Electric (which had assigned two employees to NEDO), personal communication, 8 November 1991.

35. Katsuo Seiki, MITI, interview, 12 November 1991.

36. Hideo Suzuki, "Kyoto Conference Keynote Speech."

37. Ibid.

38. Choy, "Japan's Energy Policy," 13.

39. *NEDO*, 7. The range and depth of Japan's commitment to research on alternative energy technology is thoroughly documented in reviews published in English by NEDO. Projects encompass improved technology for the manufacture of low-cost, highly efficient solar cells and solar thermal systems; fuel cells for producing electricity through chemical oxidation with no noise and almost no pollution; advanced battery storage systems with applications in electric vehicles; super-efficient heat pumps that operate without ozone-depleting CFCs; and super-efficient ceramic gas turbines.

The NEDO program combines highly sophisticated technical research with user testing and gradual learning from experience, always in close collaboration with manufacturers and customers for the technology. Much of its program seeks to commercialize technologies originally developed in the United States: solar cells and fuel cells were both products of the U.S. space program. The program has no equivalent in the United States. Further information concerning NEDO's energy research is available from the Japan program of the Center for Global Change, University of Maryland, College Park, Md.

40. *NEDO*, 7.

41. Choy, "Japan's Energy Policy," 15.

42. *NEDO*, 7.

43. Choy, "Japan's Energy Policy," 15.

44. Information about RITE comes from the following sources: an undated fifteen-page brochure entitled *RITE: Research Institute of Innovative Technology for the Earth*, in English and Japanese; an undated twenty-two-page brochure entitled *RITE NOW: 1991 Autumn* and bearing the numeral 1 in the upper right-hand corner, written in Japanese; Yoshiyuki Namba, managing director and secretary general of RITE, personal communication, 7 November 1991.

45. Myers, "Japan Bids," 1144–45. See also Myers, "A Technical Fix for the Greenhouse," *Science* 256 (22 May 1992): 1144. Five of RITE's suite of seven big projects have familiar goals: substitutes for the chlorofluorocarbon compounds that attack ozone, biodegradable plastics, improved bioreactors for low-energy synthesis of chemicals, production of hydrogen by bacteria, and recycling of steel scrap. But the two front-runners, sharing half of the total $28-million-a-year budget, are more unusual, aiming to use biological and chemical methods for removing carbon dioxide from industrial exhaust gases. Sixteen companies are involved in the work, including Hitachi, Asahi Glass, and Sumitomo Chemical Company, each bringing its own special area of expertise.

46. Permission from a government ministry is required to create such foundations. Permission for the establishment of RITE was granted by the Ministry for International Trade and Industry.

47. Myers, "Japan Bids," 1144–45.

48. Ibid.

49. Ibid.

50. Ibid.

51. Ibid.

52. Ibid.

53. Karel van Wolferen, *The Enigma of Japanese Power* (New York: Knopf, 1989), 398.

54. Ira C. Magaziner and Mark Patinkin, *The Silent War: Inside the Global Business Battles Shaping America's Future* (New York: Vintage Books, 1990), 363. Automobile- and boat-racing tax proceeds also benefit research and development. MITI guides the direction of these funds to various industry

associations, which then distribute the money to specific projects in rationalization, research, and export promotion. They are separate from the government's official budgets and help to finance sectors and products that are not household words in Japan. In 1978, racing subsidies to the machinery industries, for example, totaled ¥7.3 billion.

55. Ibid., 364.

56. "Japanese Plan for Global Warming Stimulates Major PV Initiatives," *Photovoltaic News* 11, no. 5 (May 1992): 3.

57. Jacobs and Harris, *Public-Private Partnerships*.

58. Ibid.

59. Ibid.

60. See President Bill Clinton, "Technology for America's Economic Growth: A New Direction to Build Economic Strength," speech delivered 22 February 1993.

61. John Zysman, "US Power, Trade and Technology," *International Affairs* 67 (January 1991): 81–106. The growing dominance of Japanese technology in global markets for high technology is accepted by experts but may not be fully appreciated by the public. As the editors of a report on the subject by the National Academy of Sciences state, "It is clear that U.S. technological preeminence has passed, that U.S. technological leadership is eroding in many fields, and that Japan will be a major competitor and collaborator in the future." T. Arrison and M. Harris, "Japan's Growing Technological Capability and Implications for the U.S. Economy: An Overview," in Arrison et al., *Japan's Growing Technological Capability*, 17. The issue, we are told, is to understand the implications and our options.

Chapter Three. The World Market and American Decline

1. Environmental Economics Associates, "The Environmental Industry in the United States" (Washington, D.C.: U.S. Environmental Protection Agency, January 1991). EPA's estimate of environmental expenditures is published in its "Cost of a Clean Environment" reports to Congress. Estimates of current environmental expenditures are based on detailed Census Bureau surveys of expenditures by state and local governments and manufacturing establishments. Estimates of future expenditures are based on projections of current program trends, and those for new programs on estimates from EPA's regulatory impact analyses.

2. Roger H. Bezdek, "Jobs and the Environment," *Environment*, September 1993, 7–32. See also Joel Makower, *The E-Factor: The Bottom Line Approach to Environmentally Responsible Business* (Tilden Press, 1993), 65–67. Indeed, there seems to be a correlation between profitability and environ-

mental performance. Two economics professors at Dickenson College in Pennsylvania, Stephen E. Erfle and Michael J. Fratantuono, compared standard measures of corporate performance for eighty-four companies with their ratings in the book *Shopping for a Better World*. They found a clear link between profits and social responsibility. Companies that were the most highly rated in terms of environmental protection, charitable giving, community action, and advancement of women and minorities scored as much as 16.7 percent higher in various measures of profitability than the firms rated lowest by these criteria.

3. Environmental Economics Associates, "The Environmental Industry." Spending is included, however, if it is reported in municipal or private solid-waste budgets. EPA estimates expenditures by program area (air, water, etc.), by funding source (federal, local government, private enterprise), and by type of regulation (existing, new). Estimates are provided for capital costs, operating costs, and annualized costs. Annualized costs in any one year are the operating costs for that year plus an amortization factor applied to the estimated capital stock in place. Although these annualized costs are an appropriate measure of the cost of environmental regulations, they are not a measure of the market for environmental goods and services. This market is more accurately measured as the sum of the annual capital expenditures plus the annual operating costs. This figure provides an estimate of the capital goods (plants and equipment) purchased each year for environmental purposes plus the labor, materials, and energy used to operate the capital equipment already in place. This is what the environmental industry sells—buildings and equipment, services, and materials.

4. Ibid.

5. Ibid.

6. Ibid. See also *The Environmental Industry, Open for Business: A Review of Emerging Trends and Economic Opportunities* (Eugene, Oregon: Economic Research Associates, April 1991). Unlike some other businesses, the environmental industry offers the prospect of continued growth in well-paying jobs. An Oregon study found that (1) the average 1990 pay of workers in the environmental industry was $29,400, which was about 30 percent higher than that of industry statewide; (2) the expected employment growth rate through 1996 was 4.5 percent, significantly higher than that of most other industries; and (3) the projected average national sales growth for the industry through the year 2000 was roughly 4.0 percent annually.

7. See Democratic Policy Committee, "DPC Special Report: Jobs and the Environment," publication SR-53-Environment, 2 October 1992. Projections have included the following: the Organization for Economic Co-

operation and Development (OECD) has reported that environmental goods and services is a $200-billion industry that will experience 5.5 percent growth on an annual basis; the Office of Technology Assessment (OTA) estimates that global environmental goods and services will grow to a $300-billion industry by the year 2000; and the International Finance Corporation (IFC) reported that the worldwide market for environmental goods and services is expected to grow rapidly during the next decade, doubling from roughly $300 billion to $600 billion by the year 2000.

8. "Singapore's Green Crusade: Cleaning the Neighbours," *Economist*, 1 February 1992, 80.

9. Ibid. The International Finance Corporation, in a report entitled "Investing in the Environment" (May 1992), cited specific countries with needs for environmental technology: *Thailand.* The total private sector market for pollution control equipment is estimated to be about $210 million per year. In the next decade, demand is expected to grow 20 to 25 percent. By the year 2000, the private sector market for pollution control equipment and services is expected to reach at least $1.5 billion. *Malaysia.* The combined public and private sector demand for environmental goods and services is estimated at roughly $210 million per year, and is expected to grow 15 to 20 percent annually. *Poland.* In September 1990 the Ministry of Environment introduced a new ecological policy to improve the country's environment. The total cost of implementing this policy is estimated to be $1 billion in 1991, and $7 billion over the 1991 to 1995 period. Full implementation of its environmental policies would take at least thirty years and cost $260 billion, much of it going to pay for pollution cleanup equipment that could be produced in the United States.

10. "Singapore's Green Crusade."

11. Vincent Yip and Brian Fliflet, "China, Hong Kong, ASEAN Countries Are Frontier Markets," *NewsACTION*, Spring 1992, 14–15. Singapore also is home to what may be the world's most concerted program to reduce pollution by controlling traffic. Driving to the city center requires a permit; randomly placed spotters search for violations, for which fines are levied automatically. The excellent transportation network includes rail, bus, and "metro," or subway, lines. The rigid controls include even taxis, which are equipped with alarms inside the passenger compartment that sound automatically when the speed exceeds 80 kilometers per hour (50 mph).

12. Ibid.

13. Alberto Sabadell, "Energy/Environment Technology Transfer to Developing Nations Is Big Business: Will the U.S. Lose Out?" *Strategic Planning for Energy and the Environment*, 73–77.

14. Quoted in Thomas W. Lippman, "An Electrifying Opportunity:

U.S. Utilities Join Charge to Power International Markets," *Washington Post*, 20 April 1993, D1, D3.

15. Sabadell, "Energy/Environment Technology," 73–77.

16. Lippman, "An Electrifying Opportunity," D1, D3.

17. Ibid.

18. Ibid.

19. Sabadell, "Energy/Environment Technology," 73–77.

20. Ibid.

21. "The Role of Repowering in America's Power Generation Future," a report by the Office of Fossil Energy, U.S. Department of Energy, November 1987.

22. C. F. Gibbs and M. C. F. Steel, "European Opportunities for Fuel Cell Commercialization," *Journal of Power Sources* 37 (1992): 35–43.

23. Ibid.

24. See "The Role of Repowering." The U.S. Department of Energy concluded that for fossil fuel–fired power plants the technical options available to deal with the triple challenges of aging boilers, increasing electrical demand, and steadily more demanding environmental requirements were limited to life extension, retrofit of emissions control equipment, construction of new capacity, and repowering. It rejected the first three options, saying neither life extensions nor retrofits would provide needed new capacity, while new construction wasn't cost effective. With repowering, however, "the potential exists to satisfy much of the Nation's foreseeable demand for new capacity without undertaking a new and expensive wave of 'grassroots' power plant construction."

25. Ibid. A repowered coal-fired plant would retain much of its existing solids-handling equipment and virtually all of its steam cycle and electrical generating and power-conditioning hardware, although these components may be refurbished simultaneously with the repowering. Thus, repowering is part of a life-extension program as well. In some repowering schemes, a gas turbine generator would be added to reconfigure the plant as a combined cycle (combining both gas and steam turbines in a single power train). In the aggregate, these changes reduce air pollution not only because the repowering technologies are all cleaner than conventional systems, but also because they boost efficiency.

26. *Greenwire* 3, no. 8 (12 May 1993), 8.

27. Frank Bass, "Spending on Environment Still a Priority—Support Not Eroded by Tough Economy," *Houston Post*, 3 March 1992, A12.

28. Greenberg/Lake, "New Priorities for Energy Policy," polling report, Washington, D.C., 8 February 1993.

29. "Americans Favor Legislative Action, Even Higher Costs, to Protect

Environment," *Environment USA* poll (Golin/Harris Communications, Inc., press release, November 1991).

30. "Study Finds Americans Fit Into Seven Environment-Related Segments," *Environment USA* poll (Golin/Harris Communications, Inc., November 1991).

31. Debora MacKenzie, "Stormy Weather for Insurers," *Tomorrow*, April–June 1993, 32.

32. Quoted ibid.

33. See *Current Issues in Atmospheric Change: Summary and Conclusions of a Workshop October 30–31, 1986*, by the Board on Atmospheric Sciences and Climate, Commission on Physical Sciences, Mathematics, and Resources of the National Research Council (National Academy Press, 1987). Global warming is by no means the only environmental imperative pushing the world into the new era. Neither the greenhouse effect nor the other environmental threats driving public opinion and governmental action are fictitious. Scientists know that trace gases cause global warming because their effects are well established from laboratory measurements. Satellite observations clearly show that the trace gases in the Earth's atmosphere are, in fact, absorbing radiation at the predicted wavelengths. Similarly, scientists have documented that smog and other urban pollutants are more than mere irritants.

34. David Chittick, in testimony before the Senate Committee on Environment and Public Works, 23 February 1993, Washington, D.C. (emphasis in original).

35. Because of the extensive cross-border trade in auto parts with Canada, data is best analyzed as North American rather than distinctly United States or Canadian.

36. Data supplied by the international consulting firm of Mike Walsh and Associates, of Arlington, Virginia, based on statistics compiled by the Motor Vehicle Manufacturers Association.

37. Council on Competitiveness, "A Competitive Profile of the Motor Vehicle Industry" (Washington, D.C.: March 1991).

38. RCG/Hagler Bailly described the situation in a 1991 report, *The Power Generation Market*.

39. Derek Denniston, "Second Wind," *World Watch*, March–April 1993, 33–35.

40. Makower, *The E-Factor*, 213–14.

41. Global Environment Fund, *1991 Annual Report* (Washington, D.C.: 1992), 32.

42. Organization for Economic Cooperation and Development, Direc-

torate for Science Technology and Industry, "The OECD Environment Industry: Trends and Issues" (Paris: 1991).

43. Fumio Kodama, "Changing Global Perspective: Japan, the USA and the New Industrial Order," *Science and Public Policy*, December 1991, 385–92.

44. Stephen M. Meyer, "Environmentalism and Economic Prosperity: Testing the Environmental Impact Hypothesis" (Boston: Massachusetts Institute of Technology, October 1992).

45. Ibid.

46. Stephen M. Meyer, "Environmentalism and Economic Prosperity: An Update" (Boston: Massachusetts Institute of Technology, February 1993).

47. Ibid. Meyer said "any such conclusion at this point would be speculation." Still, in a later study he tested a half-dozen other possible explanations for the link between prosperity and environmental protection and found all wanting.

48. Bezdek, "Jobs and the Environment," 7–32.

Chapter Four. U.S. Policy Failures

1. Daniel Yergin, *The Prize: The Epic Quest for Oil, Money, and Power* (New York: Simon & Schuster, 1992), 97.

2. David Davis, *Energy Politics, 4th ed.* (New York: St. Martins Press, 1993).

3. Franklin Tugwell, *The Energy Crisis and the American Political Economy* (Stanford, Calif.: Stanford University Press, 1988), 57, 173.

4. Allison Thompson, Office of Employment and Unemployment Statistics, Bureau of Labor Statistics, U.S. Department of Labor, memorandum dated 5 May 1992.

5. "The Petroleum Industry Is Losing Jobs," Editorial and Special Issues Department of the American Petroleum Institute, Washington D.C., 24 September 1992.

6. Ibid.

7. Office of Technology Assessment (OTA), *Acid Rain and Transported Air Pollutants: Implications for Public Policy* (Washington, D.C.: Office of Technology Assessment, U.S. Congress, Superintendent of Documents, 1985), 142.

8. Motor Vehicle Manufacturers Association, *Facts and Figures 1992* (Washington, D.C.), 64. See also, "Senators Present New Bipartisan Budget Plan to Cut Deficit; Will Cut Spending First While Slashing New Taxes," press release, Office of Senator Jack Danforth, Washington, D.C., 20 May

1993. Danforth, however, said he opposed the Clinton plan, which was aimed at cutting the U.S. deficit through a mixture of taxes and spending cuts, as "far too dependent on higher taxes."

9. Michael Weisskopf, "Fanning a Prairie Fire," *Washington Post*, 21 May 1993, A1.

10. Citizens for a Sound Economy, *1991 Annual Report* (Washington, D.C.), 3. Citizens for a Sound Economy (CSE) describes itself as a "citizen advocacy group that promotes market-based solutions to public policy problems to achieve economic freedom and opportunity for all people." It maintains an interlocking directorate that includes members who serve on the boards of other organizations such as the libertarian CATO Institute. CSE is housed with a subsidiary organization, Citizens for a Sound Environment Action Fund, which states in documents filed with the Internal Revenue Services (IRS form 1024, filed December 27, 1990) that one of its purposes is attempting to influence legislation concerning environmental matters, although lobbying by nonprofit organizations is illegal. Although CSE claims a membership of 250,000, it declines to provide details on individual members or breakdowns of the share of its income which is attributable to corporate versus individual contributions.

11. Council on Competitiveness, *German Technology Policy: Incentive for Industrial Innovation* (Washington, D.C.: 1992).

12. Ibid.

13. Paul H. Weaver, *The Suicidal Corporation: How Big Business Fails America* (New York: Simon and Schuster, 1988), 142.

14. Ibid., 166–67.

15. *Dodge v. Ford Motor Co.*, 204 Mich. 459, 170 N.W. 668 at 671, 3 A.L.R. 413 (1919), quoted in Norman D. Lattin, *Lattin on Corporations*, 2d ed. (Mineola: Foundation Press, 1971), 211–12.

16. *A. P. Smith Mfg. Co. v. Barlow*, 13 N.J. 145, 98 A.2d 581 at 586, 1953, cited in Lattin, *Lattin on Corporations*.

17. See "Environmental News," Office of Public Affairs, U.S. Environmental Protection Agency, 27 July 1989. During the 1986–1988 period, the number of urban areas failing to meet the federal health-based standard for ozone increased from 64 to 101. (This was at least partially explained by the fact that the summer of 1988 was the third hottest in the United States since 1931; such hot dry conditions are conducive to the formation of ozone.) The official results had been widely reported unofficially beginning in late 1988.

18. Michael Specter, "Record Hot Readings in 1980s Boost Global-Warming Theory," *Washington Post*, 13 January 1990, A2.

19. See, for example, Statement of Mack McFarland on behalf of the

Chemical Manufacturers Association, Joint Hearing of the Subcommittee on Environmental Protection and Hazardous Wastes and Toxic Substances and the Committee on Environment and Public Works, U.S. Senate, 27 October 1987. "Clearly . . . chemistry contributes to ozone depletion," said McFarland.

20. *Curbing Acid Rain: Cost, Budget and Coal-Market Effects* (Washington, D.C.: Congressional Budget Office, U.S. Congress, 1986).

21. Donald Deieso, testimony before the Committee on Environment and Public Works, U.S. Senate, 23 February 1993.

22. Curtis A. Moore, "The 1990 Clean Air Act Amendments: Silk Purse or Sow's Ear," *Duke Environmental Law and Policy Forum*, June 1992, 26–58.

23. John Lancaster, "Lobby Gains Ground in Effort to Add 'Balance' to Wetlands Laws," *Washington Post*, 15 May 1991, A17.

24. Allen James, executive director of RISE, interview, 15 January 1992.

25. Jeff Taylor, vice president of the Safe Buildings Alliance, interview, 6 January 1992; Dr. Philip Landrigan, chair, Department of Community Medicine, Mt. Sinai School of Medicine, interview, 15 January 1992.

26. See Curtis A. Moore, "Corporate Wolves in Sheep's Clothing," *TDC*, March 1992, 14.

27. See Matthew L. Wald, "Pro-Coal Ad Campaign Disputes Warming Idea," *New York Times*, 8 July 1991; Tom Helland of Simmons Advertising (the firm employed by ICE to conduct the campaign), interview, 11 August 1992.

28. Wald, "Pro-Coal Ad Campaign."

29. Ibid.

30. See David H. Freedman, "Lead Researcher Confronts Accusers in Public Hearing," *Science*, 24 April 1992, p. 437.

31. Interview with a member of Senate staff deeply involved in enactment of the Clean Air Act amendments of 1990 who asked not to be identified, 29 June 1990.

32. Bill Fay, interview, July 1990.

33. Personal communication, ASAP, 3 February 1994.

34. Wayne Walker, interview, July 1990.

35. Interview with a congressional aide who asked not to be identified, July 1990; Wayne Walker, interview, July 1990.

36. Quoted in Martha Derthick and Paul J. Quirk, *The Politics of Deregulation* (Washington, D.C.: The Brookings Institution, 1985), 212.

37. Funding for programs which were supported by industry—registration of pesticides, thus allowing their sale and use, for example—were maintained or expanded. See Lawrence Mosher, "Will EPA's Budget Cuts

Make It More Efficient or Less Effective?" *National Journal* 13, 15 August 1981, 1466–69; Lawrence Mosher, "Move Over, Jim Watt, Anne Gorsuch Is the Latest Target of Environmentalists," *National Journal* 13, 24 October 1981, 1899–1902; and Lawrence Mosher, "Reagan's Environmental Federalism—Are the States Up to the Challenge?" *National Journal* 14, 30 January 1982, 184–88. For a table showing the decline in employment between 1980 and 1983 at the independent regulatory agencies, see the *Washington Post,* 22 November 1982.

38. The recipients of EPA power were the Office of Management and Budget and an ad hoc council headed by the vice president. During the Reagan administration the ad hoc council was known as the President's Task Force on Regulatory Relief, headed by then–Vice President Bush. During Bush's own administration, the council changed names to become the Council on Competitiveness, headed by Vice President Dan Quayle.

39. Ira Magaziner and Mark Patinkin, *The Silent War* (New York: Vintage, 1990), 366.

40. OTA, "A Review of Federal Efforts to Increase Energy Efficiency in Buildings," chapter 4 of *Building Energy Efficiency* (Washington, D.C.: 1992), 105.

41. Ironically, the belief that there was no national energy strategy was so widespread that the proposition was accepted without challenge by Congress, the press, and the public. For example, when Richard Gephart, the third-ranking Democrat in the U.S. House of Representatives, delivered the Democratic party's reply to a 1990 speech by President Bush, he asserted that "for a decade America has been left with no real energy policy at all." See Curtis Moore and S. David Freeman, "Kicking America's Oil Habit," *Washington Post,* 16 September 1990, B3.

42. Robert Sherrill, *The Oil Follies of 1970–1980: How the Petroleum Industry Stole the Show (and Much More Besides)* (New York: Anchor Press, 1983).

43. Thomas W. Lippman, "U.S. Tries to Influence Oil Prices, Papers Show," *Washington Post,* 21 July 1992, A1.

44. James J. MacKenzie, *Why We Need a National Energy Policy* (Washington, D.C.: World Resources Institute, 1990).

45. Ibid.

46. World Resources Institute, *World Resources 1992–93* (Washington, D.C.: 1992), 211.

Chapter Five. California Sunshine

1. See John J. Kirlin and Donald R. Winkler, eds., *California Policy Choices*, vol. 5 (Sacramento: School of Public Administration, University of

Southern California, 1989), 57. The state's technology export program was formalized in 1986 and has been in full-scale operation since September 1987.

2. See Peter Asmus and Bruce Piasecki, "State Energy Policies and Global Warming," in Kirlin and Winkler, *California Policy Choices*, vol. 5, 73. At present Texas has 5,273 MW of operational cogeneration, the highest in the nation, of which 51 percent is sold to utilities. California comes in second with 3,294 MW of cogeneration.

3. Ibid., 67.

4. See Tim Duane, "Electricity Regulation Reform," in Kirlin and Winkler, *California Policy Choices*, vol. 6 (1990), 208. The CEC was established essentially to prevent the development of unneeded power plants, but it was not given direct authority over utility resource planning. Instead, it had an independent demand-forecasting capability, with authority to adopt an official state forecast of electricity demand; centralized siting authority for all thermal power plants over 50 MW, whereby projects could be approved only if they met established need; authority to establish conservation standards and incentives to obviate future need; and a technology development function to promote conservation and renewable energy. These four functions were translated into action by the four main divisions of the CEC: Assessments, Siting and Environmental, Conservation, and Development.

5. Section 25000.1(a) of the Warren-Alquist Act.

6. See Duane, "Electricity Regulation Reform," 209–10. Unlike the California Public Utilities Commission (CPUC), the new agency (CEC) focused only on energy issues. While the CPUC mandate was primarily one of protecting ratepayers and shareholders through the establishment of "fair and equitable rates," the CEC was required to examine more than just the direct ratepayer economics of energy project development. The CEC's resource planning process focused on determining the "need" for new resources, "which, in the judgement of the commission, will reasonably balance the requirements of state and service area growth and development, the protection of the public health and safety, the preservation of environmental quality, the maintenance of a sound economy, and the conservation of resources."

7. Peter Asmus, "Diversifying California's Energy Map," *California Journal*, July 1991, 310.

8. See Duane, "Electricity Regulation Reform," 215. The new ISO-4 contract offered fixed-price payments for up to ten years of a thirty-year contract, while cogenerators (whose fuel costs could vary over time) had special options that fixed the non-fuel component of the payment calculation. Both firm and "as-available" capacity payments were offered, but

the implicit forecast of avoided costs assumed continuing escalation of fuel costs. Ratepayers would bear the burden of forecasting errors, which could be either high or low. In return they would receive qualifying facility power at fixed prices for up to ten years. After that, the ISO-4 called for payments at prevailing short-run avoided-cost rates. The contract did not include provisions for revising the terms, however, if conditions changed dramatically during the period of its availability. This proved to be a fatal flaw in design of the otherwise attractive interim measure.

9. Ibid., 212. In issuing ISO-4, CPUC was implementing a federal law, the Public Utility Regulatory Policies Act (PURPA). Enacted in 1978, PURPA established the foundations for independent, non-utility-owned generation by providing private utilities to purchase electricity generated by private companies, a far-reaching change. PURPA did not have a significant effect until the early 1980s, partly because it was not until February 1980 that the Federal Energy Regulatory Commission (FERC), which regulates all wholesale power transactions and hydroelectric project licensing, adopted regulations allowing state regulatory commissions to implement PURPA. California regulators at CPUC, most of them appointed by Governor Brown, chose to pursue implementation aggressively.

The advantages of the S0-4 contract were several. First, it had a generous energy price for non-fossil plants during the first third of their contract term (five to ten years). The capacity price was based on the cost of a combustion turbine escalated with inflation. Finally, the contract allowed for terms from fifteen to thirty years and had pricing options tailored to the needs of both fossil and non-fossil. (Jan Hamrin, "Land of the Setting Sun," *Cogeneration and Resource Recovery*, January/February 1991, 8–9.)

10. See Duane, "Electricity Regulation Reform," 213. See also Lyna Wiggins et al., "Diversification in Energy Production," in Kirlin and Winkler, *California Policy Choices*, vol. 4 (1988). Section 210 of PURPA requires utilities to purchase the power output of qualifying facilities (QFs), which are exempt from regulation as public utilities, at the utility's "avoided cost"—defined as what it would have cost the utility to produce the power in the absence of the QF purchase. A QF must be a renewable energy technology (e.g., wind, solar, geothermal, hydroelectric) that is less than 80 MW in size, or a cogeneration project that meets minimum standards for efficient production of both electricity and thermal energy (e.g., steam or hot water). Figures compiled from third-quarter 1990 CPUC utility company filings show 9,160 MWs of ISO-4 capacity. More than 60 percent of these resources are renewable energy technologies.The success of the ISO-4 projects and the diversity of the resource mix can be attributed to several factors, according to Jan Hamrin: "1) There was a sufficient buy-back rate

(avoided cost) paid through capacity and energy prices to allow a variety of technologies to be cost effective; 2) There was a 'mixed portfolio' of pricing schemes which allowed some tailoring of the revenue stream (at the same present value) according to the needs of the technology; and 3) There was a 'standard contract' of terms and conditions which could accommodate a variety of technologies and was pre-approved as 'reasonable' guaranteeing the utility cost recovery and lowering the transactional costs of power sales contract negotiation" (Hamrin, "Land of the Setting Sun," 8).

11. See Duane, "Electricity Regulation Reform," 215.

12. Ibid., 216.

13. Hamrin, "Land of the Setting Sun," 8.

14. Ibid.

15. The use of solar PVs as a power source at remote homes such as hunting or vacation cabins is also burgeoning. PG & E estimates that within its service territory in Northern California, about 1,000 such systems were installed in 1990 alone, boosting the total there to 4,500.

16. *Photovoltaic News*, February 1992. In 1991, Siemens shipped 9 megawatts of photovoltaic cells, vastly outdistancing its nearest competitor, Solarex, which shipped only 5.2 megawatts.

17. Asmus, "Diversifying California's Energy Map," 305–11.

18. Ibid.

19. Ibid.

20. Quoted ibid.

21. Solar and wind energy estimates from Michael Brower, *Cool Energy: Renewable Solutions to Environmental Problems* (Cambridge, Mass.: MIT Press, 1991), 22. Brower, a physicist, is research director for the Union of Concerned Scientists.

22. See OTA, *New Electric Power Technogies: Problems and Prospects for the 1990s*, OTA-E-246 (Washington, D.C.: U.S. Congress, Office of Technology Assessment, 1985). While still a very small part of total energy supply, the exploitation of geothermal resources is spreading through developing nations in Latin America, Asia, and Africa; see OTA, *Energy in Developing Countries*, OTA-E-486 (Washington, D.C.: U.S. Congress, Office of Technology Assessment, 1991).

23. For energy resource estimates, see *The Potential of Renewable Energy: An Interlaboratory White Paper*, prepared for the Office of Policy, Planning, and Analysis, U.S. Department of Energy (prepared by Idaho National Engineering Laboratory, Los Alamos National Laboratory, Oak Ridge National Laboratory, Sandia National Laboratories, Solar Energy Research Institute), SERI/TP-260–3674, March 1990.

24. Brower, *Cool Energy*, 22–23.

25. See *The Potential of Renewable Energy: An Interlaboratory White Paper.* Solar thermal systems use concentrated sunlight to generate heat for thermal conversion processes, such as electricity generation. Three types of solar thermal technologies—parabolic trough systems, central-receiver plants, and parabolic dish systems—are either currently in use or under development.

Worldwide interest in solar thermal hybrid systems has increased recently. Plants are planned for India, Jordan, and Israel, and aggressive R, D & D is continuing in Spain, Germany, and Israel to capitalize on these emerging markets, lending the industry a multinational character. The major solar thermal hybrid supplier has both United States and foreign involvements. The U.S. industry hopes to expand to other high-insolation areas of the world.

26. Ibid., 8. The system operated by Luz uses mirrored troughs to focus the sun's rays on fluid-filled tubes. A second major solar thermal technology is the central-receiver plant. A 10-MW central-receiver power plant was deployed by a joint government/industry team and operated successfully for several years in a grid-connected mode by Southern California Edison Company. Thermal storage in the system would move the technology to the dispatchable category for a utility. Six hours of storage is expected to provide daytime dispatching under variable weather conditions. The average levelized capital and operating costs projected for solar thermal central-receiver stations range between 8¢ and 12¢ per kilowatt-hour for early plants, based on current component and system designs.

Finally, prototype parabolic dish electric systems, totaling about 5 MW, have been operated in a utility setting in Georgia and in Southern California. Prototype dishes with small Stirling heat engines and generators mounted at the focal point of the dish have led to significant increases in system performance and hold the world record for system conversion efficiency from sunlight to electricity (29 percent). The Stirling engine configurations may be most appropriate for small, stand-alone applications. U.S. industry involvement in this technology is beginning to increase as the technology approaches cost competitiveness in early markets. Germany, Japan, and Spain are also working on small dish system concepts for export.

27. Ibid., 9. Perhaps 20 MW of solar cells are deployed in thousands of remote, stand-alone applications. Bulk power applications of PV are currently limited. Three megawatt-scale plants were installed in the early 1980s that continue to operate reliably. Electric utilities are currently investigating potential uses for PV systems ranging from distribution system

applications to bulk power generation. A recent survey conducted by the Electric Power Research Institute (EPRI) identified 219 grid-connected PV systems with a total combined peak power rating of 11.6 MW (including 9.4 MW for the three large plants). The reliable performance of solar cells in space has established PV as a dependable technology. Costs have come down dramatically since the first solar cells were deployed for space applications, opening up a terrestrial market of approximately 42 MW per year (1989). This market has three major segments: consumer products, remote power, and bulk power generation.

28. Ibid. Under baseline assumptions, smaller high-value applications should begin about 1995, followed by major market entry for peak power supply beginning about 2005. Achieving costs of 4¢ to 7¢ per kilowatt-hour in the 2010 to 2030 period should increase penetration significantly.

The private PV industry is a mix of some forty firms, 90 percent of which are small to medium-sized firms involved in manufacturing, distribution, or service. In general, the industry, especially manufacturing, has not been profitable. U.S. companies, which were world leaders in the early 1980s, have encountered serious competition from foreign-owned companies. The U.S. PV market share has severely declined, from 65 percent in 1981 to 35 percent in 1989. Although the market for PV is growing at a 20 percent annual rate, and industry margins are improving, this economic turnaround has yet to be reflected in the United States.

29. Kendall and Nadis, *Energy Strategies: Toward a Solar Future* (Cambridge: Ballinger, 1980). See also Brower, *Cool Energy*, 72–74; and D. L. Elliot, L. L. Wendell, and G. L. Gower, *An Assessment of the Available Windy Land Area and Wind Energy Potential in the Contiguous United States* (Richland, Wash.: Pacific Northwest Laboratory, 1991). Two analysts, Kendall and Nadis, found that the windiest areas of the United States could support enough wind power capacity to provide 18 to 53 percent of the electricity consumed in 1990. The lower figure represents the most severe assumptions of land-use exclusion (that is, all land excluded from development except 90 percent of rangeland), and the upper figure represents no exclusions at all. Other areas, distributed much more widely around the country, could support even greater wind power capacity. The study found that the total wind generation potential in these areas ranges from 1.7 to 6 times current U.S. electricity demand. Only some areas are economically suitable for wind power development today (barring the inclusion of environmental costs in utility planning), but other areas may soon be if advanced wind turbines under development meet expectations.

30. *The Potential of Renewable Energy*, 9.

Chapter Six. Wheels

1. Joseph B. White, "GM Shelves Programs to Make Its Own Electric Car and Joins Ford, Chrysler," *Wall Street Journal*, 14 December 1992, A3.

2. Ibid.

3. Tom Kenworthy, "States Poised to Tighten Auto Emission Rules," *Washington Post*, 7 January 1993, A3.

4. Samuel A. Leonard, Director Automotive Emission Control, General Motors Research and Environmental Staff, letter to Michael J. Bradley, Executive Director, Northeast States for Coordinated Air Use Management, dated 8 September 1992.

5. Thomas Fitzgerald, "California Emissions Law Could Be Costly for East," *The Times*, Trenton, N.J., 1 December 1991, A1.

6. John S. Day, "Fuel Group's Study Says New Emission Rules Would Hurt Maine's Economy," *Bangor Daily News*, 7–8 December 1991, 1.

7. "Honda's Formula 1 Engineers Will Move on to Electric Vehicles," *Alternative Energy Network Online Today*, 10 February 1993.

8. Hidenaka Kato and Masato Ishizawa, "Honda Faces R & D Crossroads," *Nikkei Weekly*, 1 August 1992, 1.

9. Stuart F. Brown, "The Theme Is Green," *Popular Science*, February 1992, 50.

10. Ibid.

11. Of the six air pollutants that are commonly regulated, all but one, sulfur dioxide, are emitted in substantial amounts from tailpipes. Similarly, tailpipes account for one third or more of carbon dioxide, the principal global-warming gas. Ozone-destroying chlorofluorocarbons are emitted in large quantities by leaky automotive air conditioners and CFC-blown foams used for seat cushions. Except for occasional hot spots associated with secondary lead smelters, lead derives almost exclusively from the gasoline burned by motor vehicles. Depending on the region, 40 to 60 percent of the oxides of nitrogen and 30 to 40 percent of the hydrocarbons, the precursors of ozone, originate with motor vehicles. Between 65 and 82 percent of carbon monoxide comes from cars, trucks, and buses.

12. Michael Walsh, testimony before the Subcommittee on Clean Air and Nuclear Regulation, Committee on Environment and Public Works, U.S. Senate, 27 April 1993.

13. Ibid.

14. Ibid.

15. Quoted in Douglas Cogan, *The Greenhouse Gambit: Business and Investment Responses to Climate Change* (Washington, D.C.: Investor Responsi-

bility Research Center), 235, citing *A Look Ahead: Year 2020*, proceedings of the Conference on Long-Range Trends for the Nation's Highway and Public Transit Systems, special report 220 (Washington, D.C.: National Research Council, Transportation Research Board, 1988).

16. Cogan, *The Greenhouse Gambit*, 235, citing *Motor Vehicle Facts and Figures '91* (Detroit: Motor Vehicle Manufacturers Association, 1991).

17. Letter from Edward N. Cole, President of General Motors, to Senator Edmund S. Muskie (D-ME), 17 September 1970, reprinted in Senate Committee on Public Works, *Legislative History of the Clean Air Act Amendments of 1970*, 93rd Cong., 2nd Sess., 1974, Committee Print (Washington, D.C.: U.S. Government Printing Office, 1974), 358. Congress rejected these arguments, choosing instead to force the industry to invent ways of solving their problems. Speaking as the Senate began consideration of the proposal, Muskie replied, "The first responsibility of Congress is not the making of technological or economic judgments—or even to be limited by what is or appears to be technologically or economically feasible. Our responsibility is to establish what the public interest requires to protect the health of persons. This may mean that people and industries will be asked to do what seems to be impossible at the present time. But if health is to be protected, these challenges must be met. I am convinced they can be met."

18. David L. Greene, Daniel Sperling, and Barry McNutt, "Transportation Energy to the Year 2020," in *A Look Ahead: Year 2020*, 216.

19. Marc Ross, Marc Ledbetter, and An Feng, "Options for Reducing Oil Use by Light Vehicles: An Analysis of Technologies and Policy," American Council for an Energy-Efficient Economy, December 1991. Indeed, there was some reduction in the ratio of maximum power to weight, although almost none in interior volume, in the early 1980s. By the mid- and late 1980s, however, the manufacturers were achieving the mandated standards with vehicles of interior volume and maximum power equal to and higher than those of the early 1970s. The CAFE standards were thus an important example of successful "technology forcing" by regulation.

20. Deborah Lynn Bleviss, *The New Oil Crisis and Fuel Economy Technologies: Preparing the Light Transportation Industry for the 1990s* (New York: Quorum Books, 1988), 132.

21. Cogan, *The Greenhouse Gambit*, 266, citing Rich Taylor, "Auto Tech '90," advertising supplement to the *New York Times Magazine*, 13 May 1990.

22. The average price of gasoline in the United States rose by only 19¢ between 1978 and 1987, to about 82¢ per gallon—the lowest cost among the world's industrialized nations. Elsewhere, fuel ranges from $1.41 per gallon in Australia to $3.71 in Italy.

23. Cogan, *The Greenhouse Gambit*, 266, citing Dorin P. Levin, "New Japan Car Weapon: A 'Little Engine That Could,'" *New York Times*, 26 November 1989; U.S. Congress, Office of Technology Assessment, *Improving Automobile Fuel Economy: New Standards, New Approaches*, OTA-E-504 (Washington, D.C.: U.S. Government Printing Office, October 1991); and David Woodruff, "Detroit's Big Worry for the 1990s: The Greenhouse Effect," *Business Week*, 4 September 1989. TRW Inc. is a maker of engine valves.

24. Richard M. Nixon, "The President's Message to the Congress Recommending a 37-point Administrative and Legislative Program," 10 February 1970, reprinted in *A Legislative History of the Clean Air Act Amendments of 1970*.

25. Graham Hagey, personal communication, 9 September 1992.

26. See Curtis A. Moore, "The 1990 Clean Air Act Amendments: Silk Purse or Sow's Ear?" *Duke Environmental Law and Policy Forum*, 1992, table 2 and text.

27. Jack Smith, Southern California Gas Company, personal communication.

28. James McKenzie, World Resources Institute, Washington, D.C., personal communication, 5 June 1992. If the ZEV requirement remains in place and is adopted by the eleven Northeastern States, there could be about 2 million ZEVs on the road in the year 2003. Assuming further that each car is driven the current U.S. average of thirty miles per day, consuming 0.5 kilowatt-hours per mile (the consumption of Chrysler's TEVan, the most power-hungry of the EVs developed so far), each car would consume 15 kilowatts during an overnight, eight-hour recharge. The aggregate consumption of this EV fleet would be about 4 million kilowatts, or a 1 percent increase in peak demand. Converting the entire fleet to battery operation would increase electricity requirements by roughly 25 percent—but *decrease* carbon dioxide emissions by a like amount, assuming the current mix in electricity consumption.

29. Pandit Patil, U.S. Department of Energy, Mobile Fuel Cell Program, personal communication, 5 June 1992. Fuel cells enjoy several advantages over batteries. Fuel cells can operate on a variety of fuels, ranging from methanol to hydrogen, for those circumstances where electricity is unavailable or undesirable. They can power heavy-duty vehicles—buses, trucks, and even locomotives. And the relative efficiency of current fuel cells (that is, the percentage of original energy which is ultimately put to a useful purpose) is currently about 32 percent, compared to about 12 percent for batteries.

30. Paul Howard, Ballard Industries, personal communication, March

1993. One well-known energy analyst, Dr. Robert Williams of the Princeton University Center for Energy and Environmental Studies, is exceptionally enthusiastic about the PEM fuel cell, extolling it as having "an order of magnitude higher power density . . . simple construction . . . no toxic or dangerous materials and no exotic materials other than the platinum . . . [and] readily amenable to dramatic cost reductions."

31. Michael Walsh, testimony before the Subcommittee on Clean Air and Nuclear Regulation, Committee on Environment and Public Works, U.S. Senate, 27 April 1993. EPA has also tested a two-stroke engine manufactured by Orbital and found it to have exceptional fuel economy while virtually meeting California's ultra-low-emission level requirements.

32. Ibid.

33. Ross, Ledbetter, and Feng, "Options for Reducing Oil Use by Light Vehicles." When the U.S. Department of Transportation's National Highway Traffic Safety Administration crashed 1984 to 1988 model year cars into a fixed barrier at 35 miles per hour, there was no relationship between automobile weight and head injury, important in itself, but also a measure of personal injury generally. In fact, there were some heavy vehicles that performed poorly and some light vehicles that performed very well.

The U.S. Department of Transportation's Research Safety Vehicle Program, which existed from 1977 to 1980, developed an experimental car that was both safe and fuel efficient. The program concluded that a car using 1980 technology (and even better materials and techniques have since been developed) could carry five passengers, achieve 43 mpg, and withstand 80-mph frontal impacts, 50 mph side impacts, and 45 mph rear impacts.

34. The heat created by combustion—even the flame of a match—causes the oxygen and nitrogen found in ambient air to react with one another to form oxides of nitrogen. Thus, a conventional internal combustion engine fueled with hydrogen creates pollution not because of the fuel, but due to its flame. A cooler flame, or perhaps using some hydrogen to chemically reverse the process, could theoretically lead to zero pollution.

35. Cogan, *The Greenhouse Gambit*, 249, citing *Motor Vehicle Facts and Figures '91*.

36. *Motor Vehicle Facts and Figures '91*. Sales of domestic cars within the United States have dropped from 9.3 million cars in 1978 to 6.9 million in 1990. From 1981 to 1986, the Big Three lost the U.S. market at an average rate of .46 percent per year. In the period between 1986 and 1990, however, that rate more than tripled, to an average of 1.5 percent per year. During this same period, the definition of "domestic" changed radically. Under intense pressure, the Japanese voluntarily restricted the number of

cars they would export to the United States and instead built factories here. Today, Honda, Toyota, Nissan, Mazda, and the other Japanese companies make one of every seven so-called "domestic" cars at their captive U.S. plants.

37. Cogan, *The Greenhouse Gambit*, 249–50, citing Paul Ingrassia, "Auto Industry in U.S. Is Sliding Relentlessly into Japanese Hands," *Wall Street Journal*, 16 February 1990.

Chapter Seven. Clean Power Technologies and Cleaner Fuels

1. *The Role of Repowering in America's Power Generation Future*, a report by the Office of Fossil Energy, U.S. Department of Energy, November 1987.
2. Ibid.
3. Ibid.
4. D. F. Spencer et al., "The Cool Water IGCC Program: Status and Results" (Palo Alto: Electric Power Research Institute, undated, but apparently 1985), table 9. Cool Water stack emissions, burning a Utah coal, were as follows: NO_x—.059 lbs per Mbtu; SO_2—.034 lbs per Mbtu; TSP—.001 lbs per Mbtu.
5. Ian M. Torrens, "Global Greenhouse Warming: Role of Power Generation Sector and Mitigation Strategies," in *Energy Technologies for Reducing Emissions of Greenhouse Gases*, proceedings of experts' seminar (Paris: Organization for Economic Cooperation and Development, 1989), vol. 2, 15–52. Torrens was with the Electric Power Research Institute, the trade association of U.S. electric power companies.
6. Michael T. Burr, "A Solid Fuel Future," *Independent Energy*, October 1992, 24.
7. Scott Gawlicki, "Selling Gasification," *Independent Energy*, March 1991, 30–38.
8. Ibid.
9. William Ellison, William Ellison and Associates, personal communication, 17 May 1993.
10. I. J. Graham-Bryce, "Optimization of Energy Strategies for Power Generation in Relation to Global Climate Change," in *Energy Technologies for Reducing Emissions of Greenhouse Gases*, vol. 2, 159–72. Graham-Bryce was with Shell International.
11. Masahashi Hatano, "Coal Utilization Technologies on Japanese Electric Power Companies," in *Energy Technologies for Reducing Emissions of Greenhouse Gases*, vol. 1, 339–60. Hatano is an officer of the Electric Power Development Company.

12. Krishna K. Pillai, "Repowering with PFBC for Efficiency, Emissions and Heat Rate Improvement," presented at the Seminar on Fluidized-Bed Combustion Technology for Utility Applications, Palo Alto, California, Electric Power Research Institute, 3–5 May 1988.

13. Krishna K. Pillai, "Pressurized Fluidised Bed Combustion," in *Electricity: Efficient End-Use and New Generation Technologies and Their Planning Implications,* Thomas B. Johansson, et al., eds. (Lund, Sweden: Lund University Press, 1989).

14. Ibid.

15. Scott M. Gawlicki, "PFB — The Next Step," *Independent Energy*, February 1992, 15–19.

16. Pillai, "Pressurized Fluidised Bed Combustion." The groundwork was laid for the Malmo demonstration by experimental units operated in England at Grimethorpe—again, paid for in part by the U.S. government.

17. Thomas H. Lee, Ben C. Ball, Jr., and Richards Tabors, *Energy Aftermath* (Boston: Harvard Business School Press, 1990), 81–82. Turbines could still be ordered if a turbine were being used to supply both heat as well as electricity, or if the turbine was the first stage of a planned IGCC plant in which coal would supposedly be the fuel ultimately used. Although a permit would not be granted for a gas-fired plant, it would be for the first stage of an IGCC system. Thus, as utility desires to buy gas turbines (and to a lesser degree combined-cycle systems) mounted, so did requests for licenses to build the first part of IGCC systems. Of course, the second halves—the coal handling equipment and the gasifier—might never be added.

18. Michael T. Burr, et al., "The 1992 Industry All Stars," *Independent Energy*, September 1992, 48–62.

19. Ibid., 48–62. Carbon monoxide emissions are 0 to 3 parts per million.

20. Ibid.

21. Eugene Zeltman, interview, 10 May 1993.

22. "ABB Increases Manufacturing Capabilities," *Independent Energy*, February 1993, 44. Re Siemens, see "The Need for Speed," *Independent Energy,* May–June 1993, 24.

23. Quoted in Burr, et al., "The 1992 Industry All Stars," 48–62.

24. San Diego Gas and Electric, South Bay 3 Augmentation Project: Application for Certification, submitted to the California Energy Commission, January 1990, docket nr. 90-AFC-1, 115.

25. "Contract Updates," *Independent Energy,* March 1993, 44.

26. *National Energy Strategy*, "Technical Annex 5: Analysis of Options to

Increase Exports of U.S. Energy Technology" (Washington, D.C.: U.S. Department of Energy, 1991/92), 46.

27. Ibid.

28. Global Environment Fund, *Annual Report 1991* (Washington, D.C.), 31–32.

29. Although California's limits are easily as stringent as those in Europe and Japan, the state generates virtually no electricity from coal.

30. Gawlicki, "PFB — The Next Step," 15–19.

31. *Confronting Climate Change: Strategies for Energy Research and Development* (Washington, D.C.: National Academy Press, 1990), 27.

32. Ibid.; for figures on expenditures, see p. 26.

Chapter Eight: The Inevitable Solution

1. See Office of Technology Assessment (OTA), *New Electric Power Technologies: Problems and Prospects for the 1990s*, OTA-E-246 (Washington, D.C.: July 1985). As noted before, LUZ's trough plant is only one of several versions of solar thermal technology. Solar thermal electric plants convert radiant energy from the sun into thermal energy, a portion of which is subsequently transformed into electrical energy. Among the systems there are four that with some feasible combination of reduced costs and risks and improved performance could be deployed within the 1990s in competition with other technologies and without special and exclusive government subsidies. They are central receivers, parabolic troughs, parabolic dishes, and solar ponds.

A central receiver is characterized by a fixed receiver mounted on a tower. Solar energy is reflected from a large array of mirrors, known as heliostats, onto the receiver. Each heliostat tracks the sun on two axes. The receiver absorbs the reflected sunlight and is heated to a high temperature. Within the receiver is a medium (typically water, air, liquid metal, or molten salt) which absorbs the receiver's thermal energy and transports it away from the receiver, where it is used to drive a turbine and generator, though it first may be stored.

Parabolic dishes consist of many dish-shaped concentrators, each with a receiver mounted at the focal point. The concentrated heat may be utilized directly by a heat engine placed at the focal point (mounted-engine parabolic dish), or a fluid may be heated at the focal point and transmitted for remote use (remote-engine parabolic dish). Each dish/receiver apparatus includes a two-axis tracking device, support structures, and other equipment.

With a parabolic trough, the concentrators are curved in only one dimension, forming long troughs. The trough tracks the sun on one axis, from east to west as the sun moves across the sky. A heat-transfer medium, usually an oil at high temperature (typically 200 to 400° C), is enclosed in a tube located at the focal line. A typical installation consists of many troughs, and the oil-carrying tubes located at their focal lines are connected on each end to a network of larger pipes. Whether stored in tanks or used immediately, the oil ultimately passes through a heat exchanger where it transfers energy to a working fluid such as water or steam, which in turn is routed to a turbine generator. At the LUZ solar energy generating units (SEGs), the only large trough installations in the United States, the oil's heat is supplemented with a natural gas–fired combustion system to obtain adequate steam temperatures to drive the turbine.

2. Kenneth Sheinkopf, "Waiting for Tax Credits," *Independent Energy*, April 1992, 61–64.

3. James Caldwell, personal communication, 18 May 1993. Caldwell is a former president of ARCO Solar, which was a wholly owned subsidiary of the Atlantic Richfield Co. before its sale to Siemens. Caldwell was also a vice president of the parent company.

4. "Looking Back: The First Twenty Years," *Independent Energy*, May/June 1991, 22–25.

5. Thomas B. Johansson, et al., *Electricity: Efficient End-Use and New Generation Technologies and Their Planning Implications* (Lund, Sweden: Lund University Press, 1989), 17.

6. OTA, *Energy Technology Choices: Shaping Our Future* (Washington, D.C.: 1991), 65. About 89.8 percent of U.S. electricity is generated from fuels of which there is a finite supply: 54.9 percent from coal, 9.4 percent from natural gas, 3.9 percent from oil, and 21.7 percent from nuclear (Edison Electric Institute, *Guide to the Electric Utility Industry*, Washington, D.C., 1991, 17).

7. OTA, *Energy Technology Choices*, 25.

8. OTA, *Catching Our Breath: Next Steps for Reducing Urban Ozone* (Washington, D.C.: 1989), 19.

9. A photovoltaic cell is a thin wafer of semiconductor material which converts sunlight directly into direct current (DC) electricity by way of the photoelectric effect. Cells are grouped into modules, which are encapsulated in a protective coating. Modules may be connected to each other into panels, which then are affixed to a support structure, forming an array. The arrays may be fixed or movable, and are oriented towards the sun. Any number of arrays may be installed to produce electric power, which, after

conversion to alternating current (AC), may be fed into the electric grid. In a PV installation, all the components other than the modules themselves are collectively termed the balance-of-system.

At present, PV systems are being pursued in many different forms. Each seeks some particular combination of cost and performance for the module and for the balance-of-system. In a concentrator module, lenses are used to focus sunlight received at the module's surface onto a much smaller surface area of cells; all available concentrator systems follow the sun with two-axis tracking systems. A flat-plate module is one in which the total area of the cells used is close to the total area of sunlight hitting the exposed surface of the module. Various mechanisms such as mirrors can be used to divert light from adjacent spaces onto the exposed surface of the modules. Flat-plate systems may be fixed in position or may track the sun with either single- or two-axis tracking systems.

The parallel development of these two types of PV modules and systems reflects a basic technological problem: it is difficult to produce PV cells which are both cheap and highly efficient. Some PV systems being developed for deployment in the 1990s are emphasizing cells which are relatively cheap and inefficient; such cells are used in flat-plate modules. Others are using smaller numbers of high-cost, high-efficiency cells in concentrator systems. Flat-plate modules using amorphous silicon are expected to continue to develop, but basic technical improvements must be made. See OTA, *New Electric Power Technologies*.

10. These advantages of PV systems can be summed up as follows: zero pollution; free, secure, and renewable energy source, widely available and with no off-site, fuel-related impacts; no fuel supply infrastructure required; short lead-times; wide range of installation sizes; declining costs; relatively small water needs; little or no routine emissions; siting flexibility.

11. Among the disadvantages of the technology are: intermittent supply of energy; capital intensive; land extensive; exposure to the elements and to malevolence. There is no insurmountable technological obstacle to the widespread deployment of solar photovoltaics, if customers were willing to pay up to five times more for electricity in exchange for zero pollution. At $2 per watt, PVs might generate electricity for grid-connected central power stations or make grid systems obsolete by displacing them with dispersed power-generating units located, say, on rooftops. At higher prices, PVs will fill a range of smaller markets: *Remote installations:* solar power charges batteries on navigation aids and replaces diesel and other generators at radio relay and similar facilities. This market can be penetrated even at prices of $6 per watt. *Rural markets:* solar replaces diesel and gasoline generators, human and animal power, the grid, or even windmills

in rural locations such as farms, ranches, and vacation homes. Penetration can begin at $4 per watt and climb sharply as prices drop to $3 and $2. *Suburban markets:* as solar prices approach $2.50 per watt, the suburban market, where customer density is so low that it's difficult to support the overhead cost of the grid, opens to PVs. Commercial customers, attracted by prices that compete with the retail cost of grid electricity and reliability which is superior, begin to opt for solar. As the price reaches $1.50 per watt, this market explodes and solar begins to replace coal, oil, natural gas, and nuclear power. However, this triggers a new set of technological obstacles, because solar has not been extensively developed for baseload applications.

12. "U.S.–Japan Joint Venture Finds Solar Power's 'Holy Grail': 10% Efficiency," *Electric Power Daily*, 19 January 1994.

13. See Paul Maycock and Edward Stirewalt, *A Guide to the Photovoltaic Revolution* (Emmaus: Rodale Press, 1985); quote, p. 121.

14. The specific target was $1 per watt of installed solar generating capacity. At the 1993 cost for capital, that translates to 6¢ per kilowatt hour for sites with the highest quality sunlight and 12¢ for those with the worst.

15. Ira C. Magaziner and Mark Patinkin, *The Silent War: Inside the Global Business Battles Shaping America's Future* (New York: Vintage Books, 1990), 210.

16. Ibid.

17. Ibid., 211.

18. Much of the following information is based on an interview with Paul Maycock, former head of the DOE solar program, 29 October 1992.

19. These faults included problems in power conditioning harmonics, breakage of conductors because of temperature-induced expansion and contraction, and deterioration of coatings because of exposure to ultraviolet light.

20. James Caldwell, personal communication, 18 May 1993.

21. Maycock and Stirewalt, *A Guide to the Photovoltaic Revolution*, 116–17.

22. Ibid., 117.

23. Ibid., 117, 212–13.

24. Ibid., 118–19.

25. Magaziner and Patinkin, *The Silent War*, 199–230.

26. Quoted ibid., 216.

27. Ibid., 228.

28. "Japanese Plan for Global Warming Stimulates Major PV Initiatives," *Photovoltaic News* 11, no. 5 (May 1992): 3.

29. Ibid.

30. Ibid.

31. Ibid.

32. "USSC To Build $30 Million/Ten MW Per Year/Triple Stack Amorphous Silicon Plant in Virginia," *Photovoltaic News*, May 1993, 3.

33. Ibid.

34. "U.S.–Japan Joint Venture Finds Solar Power's 'Holy Grail.'"

35. See, for example, Joan Ogden and Robert Williams, *Solar Hydrogen: Moving Beyond Fossil Fuels* (Washington, D.C.: World Resources Institute, 1989).

36. See A. J. Appleby and F. R. Foulkes, *Fuel Cell Handbook* (New York: Van Nostrand Reinhold, 1981). Operating on hydrogen, fuel cells produce only pure water and pure electricity, no pollution whatsoever. Moreover, fuel cells should be able to convert almost all their input energy into electricity, unlike internal combustion engines, which are limited in their conversion efficiency by the Carnot cycle.

Fuel cells, like solar photovoltaic cells, were merely laboratory curiosities until the advent of the U.S. space program (like photovoltaics, they date to the nineteenth century). Two modern breakthroughs occurred in 1959; Dr. Harry Karl Ihrig of the Allis-Chalmers Manufacturing Company in the United States demonstrated a twenty-horsepower fuel cell–powered tractor, and an English researcher announced that he and his co-workers had developed, built, and demonstrated a five-kilowatt system capable of powering a welding machine, a circular saw, and a two-ton capacity forklift truck. In 1964, under contract to the Electric Boat Division of General Dynamics, Allis-Chalmers produced a 750-watt fuel cell system to power a one-man underwater research vessel. For submarines, hydrogen-based fuel cells were ideal power plants: compact, silent, zero-polluting, and highly efficient, they also produce potable water—qualities which quickly attracted the attention of the space program.

37. Paul Howard, Ballard Industries, personal communication.

38. This and related quotes are from an interview with Graham Hagey, 9 September 1992.

39. Ibid.

40. During the same period, said Hagey, the Department of Defense was supporting the development of fuel cells for remote power. But by about 1980, DOD decided to make petroleum-based diesel fuel its fuel of choice, which effectively precluded the use of fuel cells.

41. R. Anahara, Fuji Electric, personal communication, 8 November 1991.

42. R. Anahara, "A Perspective on PAFC Commercialization by Fuji Electric," presented at Second Annual Grove Fuel Cell Symposium, London, England, September 1991.

43. The Rokko Island Test Center is located on a small plot of land in Osaka. The purpose of the facility is to identify and eliminate problems created when alternative energy generating systems are connected to the utility grid. The facility is testing simulated grid hookups with solar photovoltaic, wind turbine, and fuel cell systems. The facility simulates not only the utility grid, but the homes and businesses in which the systems might be installed, even to the extent of including small "houses" which contain actual appliances such as refrigerators, air conditioners, televisions, and washing machines. At the conclusion of the six-year program, the Japanese government and its domestic manufacturers will have tested and perfected for commercial use wind, solar, and fuel cell generating systems, and their associated inverters, and will also have developed utility practices and procedures to overcome the potential dangers associated with wide-scale deployment of these systems.

44. NEDO is now officially known as the New Energy and Technology Development Organization, although it is still commonly referred to as NEDO.

45. This information was collected during site visits in Japan in 1991 and 1993.

46. Atusushi Fukutome, "Development of Fuel Cell Power Generation Technology: Phosphoric Acid Type Fuel Cells," presented at the Tenth Annual Conference on Energy Conversion and Storage Technologies, NEDO, October 1990.

47. Ibid.

48. R. Anahara, personal communication, 8 November 1991.

49. Wind turbines may be classified according to the amount of electricity they generate under specified wind conditions. A small turbine generates up to 200 kWe, an intermediate-sized turbine can deliver from 100 to 1,000 kWe, and a large system may produce more than 1 MWe.

50. Christopher Anderson, "The Windmill Gets a Second Wind," *Nature*, 5 December 1991, 344.

51. Ibid.

52. Randall Swisher, interview, May 1992.

53. Jim Moriarty, CEO of Carter Wind Systems, interview, 19 May 1993. Carter shipped ten turbines to England in 1993 to be installed in Cambria, which is in the north of England, near the coast and the border with Scotland.

54. Paul Gipe, "Windpower in the U.K.," *Independent Energy*, April 1991, 60–62.

55. Randall Swisher, interview, May 1992.

56. American Wind Energy Association, "European Wind Energy Incentives," Washington, D.C., 19 February 1992.

57. Ibid.

58. Ibid.

59. Ibid.

60. But in other instances the avoided cost can be zero—that's right, *zero*. The theory is that in some circumstances (e.g., nuclear power plants) a utility power plant can't be throttled back. It must, for technical reasons, operate at full speed or not at all. Thus, even if 100 megawatts were added to the system, the output of the plant—and, hence, its fuel consumption, the labor costs, etc.—would remain unchanged. Since the utility saves zero, the avoided cost is zero.

61. OTA, *New Electric Power Technologies*.

62. Anderson, "The Windmill Gets a Second Wind," 344.

63. OTA, *New Electric Power Technologies*.

64. *Energy Technology R & D: What Could Make a Difference?* 2, Oak Ridge National Laboratory, 1989, part 2, 146.

65. Jon Choy, "Japan's Energy Policy: A Powerful Example?" Japan Economic Institute report, 25 June 1993 (Washington, D.C.).

66. Ibid.

67. World Resources Institute and International Institute for Environment and Development, *World Resources 1986*, 110.

68. *Commercial Nuclear Power: Prospects for the United States and the Rest of the World* (Washington, D.C.: U.S. Energy Information Administration, 1985); and *World Resources 1986*, 109. Prior to this recent slowdown, however, nuclear power's contribution increased dramatically, from 62.4 million tons of oil equivalent in 1974 to 282.2 in 1984. In 1984, nuclear power accounted for 3.9 percent of world primary energy consumption and 16 percent of electricity produced in the OECD countries. More than 90 percent of nuclear electricity production is in highly industrialized countries, primarily in North America (35.9 percent) and Western Europe (37.1 percent).

Chapter Nine: Green Prophets

1. Except as otherwise noted, the quotations from Jeffrey Leonard and other information concerning the Global Environment Fund were provided during an interview with Leonard in April 1993.

2. Jeffrey Leonard, "An Investor Rates Best Opportunities," *In Business*, January/February 1992, 56.

3. Global Environment Fund, *1991 Annual Report*. The Global Environ-

ment Fund describes itself as "an investment partnership whose primary objective is to realize long-term capital appreciation through investments that promote environmental improvement. The Fund is designed for institutional and qualified investors seeking to benefit from increased worldwide demand for pollution control services, waste minimization techniques, advanced resource recovery processes, improved natural resource management systems, and cleaner and more efficient methods of producing and using energy."

4. Ibid.

5. Global Environment Fund, *1992 Annual Report.*

6. Except as otherwise noted, quotations from officials of the Southern California Gas Company and other information regarding the firm are based on interviews conducted at the corporate headquarters in Los Angeles in April 1993.

7. Southern California Gas Company, "Natural Gas Vehicle Program Fact Sheet," May 1992.

8. "Ford Announces Developments Leading to Natural Gas Truck Production," *Automotive Wire*, 1 October 1991.

9. "Southern California Gas Co.–SCAQMD Unveil First Commercial Fuel Cell in U.S. in Landmark Air Quality Achievement" (Diamond Bar, Calif.: South Coast Air Quality Management District, 22 May 1992).

10. Ibid.

11. *An Alternative Energy Future* (Washington, D.C.: The Alliance to Save Energy, American Gas Association, Solar Energy Industries Association, April 1992).

12. "A Healthy Balance: AT & T Environment and Safety Report 1992" (Basking Ridge, N.J.: AT & T, 1993), 6–7.

13. Unless otherwise noted, quotations from officers of AT & T and other information regarding the company are based on interviews conducted during April 1993 at the corporate headquarters in Basking Ridge, N.J.

14. "A Healthy Balance," 2.

15. Ibid., 1.

16. Ibid.

Chapter Ten. Facing the Future

1. Christopher T. Hill, "New Manufacturing Paradigms—New Manufacturing Policies?" *The Bridge*, Summer 1991, 15–24.

2. Theodore Sorenson, *Kennedy* (New York: Harper & Row, 1965), 526.

3. *World Resources 1992–93*, a report by the World Resources Institute in

collaboration with the United Nations Environment Programme and the United Nations Development Programme, 21–23. A primary reason for the more wasteful use of gasoline in the United States is that the average price is extraordinarily low by industrial-world standards. In 1990, gasoline prices hovered around 30¢ per liter in the United States; in other OECD countries they ranged from above 50¢ per liter in Germany to about $1 per liter in Italy. The fleet efficiencies of cars and light vehicles show a similar distribution, ranging from the United States (least efficient) to Italy (most efficient). Clearly, higher fuel prices and more efficient energy use go hand in hand.

4. Office of Technology Assessment, *Energy Technology Choices: Shaping Our Future* (Washington, D.C.: U.S. Government Printing Office, 1991), 65. In 1980, electricity accounted for about 10 percent of final energy demand in the world, with most analysts projecting that fraction to increase (Thomas B. Johansson, et al., *Electricity: Efficient End-Use and New Generation Technologies, and Their Planning Implications* [Lund, Sweden: Lund University Press, 1989], 17).

5. Edison Electric Institute, *Guide to the Electric Utility Industry* (Washington, D.C.: 1991), 17.

6. An excellent environmental reason for targeting cars is the astronomical increase in total motor vehicle fuel use in the United States. It's risen nearly 40 percent since 1970. Moreover, under presidents Bush and Reagan, new car efficiency *dropped* by 4 percent between 1988 and 1990, from 28.6 mpg to 27.4. Recapturing that lost one mile per gallon would be the carbon dioxide equivalent of shutting down six coal-fired power plants.

7. Council on Competitiveness, *Gaining New Ground: Technology Priorities for American's Future* (Washington, D.C.: 1992), 57.

8. Curtis Moore and David Freeman, "Kicking America's Oil Habit: We've Got the Technology—What We Need Is the Willpower," *Washington Post*, 16 September 1990, B3.

9. Adam Smith, quoted by Andrew Skinner in the preface to Adam Smith, *The Wealth of Nations*, books 1–3 (Penguin Books, 1986), 26.

10. Ibid., 28.

11. James M. Lents and William J. Kelly, "Clearing the Air in Los Angeles," *Scientific American*, October 1993, 32–39.

Epilogue

1. Quoted in Ruth Marcus, "U.S. to Sign Earth Pact; Clinton Also Backs Emissions Target," *Washington Post*, 22 April 1993, A1.

2. Matthew L. Wald, "Government Dream Car," *New York Times*, 30 September 1993, A1.

3. Quoted in Peter Behr, "White House and Business Join Hands," *Washington Post*, 30 September 1993, A1.

4. Ibid.

5. "Historic Partnership Forged with Auto Makers Aims for 3-Fold Increase in Fuel Efficiency in as Soon as Ten Years," White House press release, 29 September 1993.

6. Wald, "Government Dream Car."

7. Quoted in Eliot Marshall, "Reinventing the Automobile—And Government R & D," *Science*, 8 October 1993, 172.

8. Quoted in Warren Brown and Frank Swoboda, "Carmakers Seek Trade-Off on Electric Car Plan," *Washington Post*, 20 October 1993, A1.

9. Quoted in Nathaniel C. Nash, "How a Huntsman Stalks His Legislative Prey," *New York Times*, 4 June 1986.

10. Margaret Kriz, "Lukewarm," *National Journal*, 14 August 1993, 2028–31.

11. Gary Lee, "Clinton Sets Plan to Cut Emissions," *Washington Post*, 18 October 1993, A1. Some environmental groups were unhappy. Clinton's proposals, said a representative of Greenpeace, were merely "a repackaging of some old ideas and a few scattered new ones. There is no guarantee that any of this will get us anywhere. The only thing worse than this policy would have been no policy."

12. Maria L. LaGanga, "State Audit Calls AQMD Lax on Smog Control," *Los Angeles Times*, 27 July 1993, A1.

13. Andrew LePage, "Voters Angry Over Tax Oust Board," *Los Angeles Times*, 14 July 1993, B1.

14. Eric Bailey, "O. C. Voice on Smog Panel Is Target of Bill," *Los Angeles Times*, 12 September 1992, A1.

15. Associated Press, "Governor Removes Clean-Air Board Chief," *Daily Breeze*, 18 November 1993, B7.

16. Marcos Breton, "SMUD's Chief Accepts Top Job at N. Y. Utility," *Sacramento Bee*, 31 January 1994, A1.

17. Personal communications from numerous PG & E executives, December 1993.

18. Howard Fine, "Reformer Brown: CEQA Bill Shows Sacramento's Shift Toward Business," *Orange County Business Journal* "Special Report," 20 September 1993.

19. Donald W. Nauss and Michael Parrish, "Big 3 Try to Put Brakes on Push for Electric Cars," *Los Angeles Times*, 25 October 1993, A1.

20. "Report on Regulatory Issues, Meeting of the Board of Directors, American Automobile Association, October 19, 1993." Evidently left unintentionally at a public telephone, the memorandum describes the U.S. auto industry's campaign to simultaneously turn back the California emissions standards in the Northeast, Washington, D.C., and California itself. The document recommends that AAMA develop "a broad, consistent and joint strategic plan to address California-related issues." The campaign would include several elements, including claims "that battery technology may not permit a consumer-acceptable EV," and options designed to obtain "relief in California through all appropriate channels (CARB, Governor Wilson, California Legislature, Vice President Gore, FIP [Federal Implementation Plan], etc.)."

Index